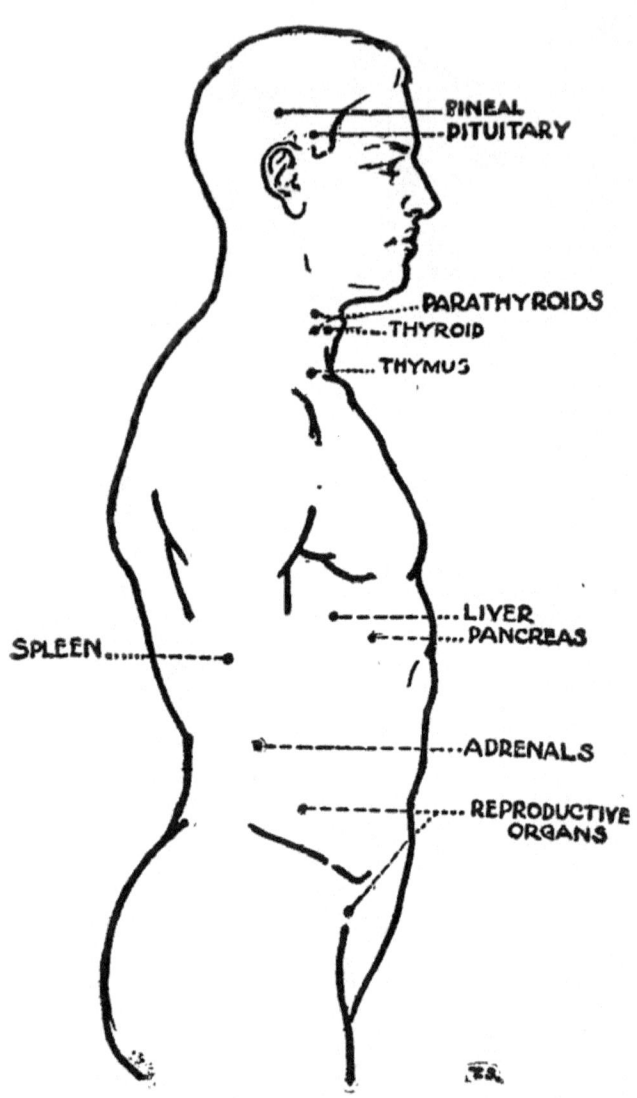

PINEAL
PITUITARY
PARATHYROIDS
THYROID
THYMUS
LIVER
PANCREAS
SPLEEN
ADRENALS
REPRODUCTIVE
ORGANS

THE DUCTLESS GLANDS

Glands in Health and Disease (1922)

By
Benjamin Harrow

©1922

a r V
18257

A 515590

Sage

To

Professor WILLIAM J. GIES
TEACHER AND FRIEND

PREFACE

Monkey glands; clever men and stupid ones; glands as the cause of crime; the origin of races; the mentally unbalanced; many acute diseases; "the bearded lady, the giant, the fat boy and the midget"; —all these and more have been dealt with under the subject of glands of internal secretion (also called ductless or endocrine glands). As in any subject that fires the popular imagination, fact and fancy have been mixed,—several drops of fact have been largely diluted with many drops of fancy. The achievements, judged by rigid scientific standards, are no more than modest, but the possibilities are limitless. It is because of these vast possibilities that an imagination, not sufficiently tempered by self-criticism, is apt to enlarge a molehill into a mountain.

The glands discussed in this book, the ductless or endocrine glands, regulate the activities of the organism little less than does the brain itself. Let but one of these bodies stop functioning, or let it function imperfectly, and injuries to various parts of the body become manifest. Since glandular disorders are among the commonest causes of disease, ranging from slight mental or physical derangement to complete mental or physical breakdown, a general survey of the subject, told in popular but not sensational language, faithfully recording what has been accomplished, should prove of interest. There is a crying need, in the opinion of the author, of simple, yet clear and clean-cut statements of scientific work to which the layman can refer. The half-baked knowledge that he often gets at present is worse than no knowledge at all.

Not the least interesting part of this fascinating field of research is the evidence we possess that the activities of these glands are dependent upon relatively simple chemical substances contained in them, to which the name "hormones" or "chemical messengers" has been given. Without in any way attempting to exaggerate their importance, we may say of these hormones that they are as indispensable to the life process as are vitamines, a subject treated by the author in an earlier volume.

I am indebted in various ways to the following: Professor W. B. Cannon (Harvard); Professor W. J. Gies (Columbia); Dr. Max Kahn (Columbia); Dr. E. C. Kendall (Mayo Clinic, Rochester, Minn.); Dr. E. G. Miller, Jr. (Columbia); Mr. A. L. Robert; Mr. Thomas Spector; Professor G. N. Stewart

(Western Reserve Univ.); and Mrs. N. J. Waller-
stein. Dr. Kendall, Prof. Stewart and Mrs. Wal-
lerstein have been kind enough to read the manu-
script and to offer several helpful suggestions.

BENJAMIN HARROW.

CONTENTS

GLANDS
IN
HEALTH AND DISEASE

"Fifty years ago the thyroid, the pituitary body, and the suprarenal capsules were mere names. Little was known of their structure, nothing of their functions. The account which we are now able to give of these organs reads like a fairy-tale. That one of the smallest should by its secretion be able to influence the growth and stature of the body, rendering this man a giant, that man a dwarf; that another should produce a material without which the nervous system is not in a condition to perform its functions; that yet others should elaborate materials which when discharged into the blood exercise a profound influence upon the activity of totally distinct and distant organs of the body, are secrets of Nature which were unrevealed fifty years ago. . . ." SIR E. A. SCHAFER, Professor of Physiology, Edinburgh University.

"More and more are we forced to realize that the general form and external appearance of the human body depends, to a large extent, upon the functioning, during the early developmental period, of the endocrine glands. Our stature, the kind of faces we have, the length of our arms and legs, the quantity and location of our fat, the amount and distribution of hair on our bodies, the tonicity of our muscles, the sound of our voice, and the size of the larynx, the emotions to which our exterior gives expression,—all are to a certain extent conditioned by the productivity of our glands of internal secretion. . . ." L. F. BARKER, Professor of Medicine, Johns Hopkins University.

GLANDS IN HEALTH AND DISEASE

CHAPTER I

INTRODUCTORY

An analogy.—The Headquarters Staff is often spoken of as the "brains" of the army. We can speak with equal justice of the brain as the Headquarters Staff of an army consisting of the millions of cells of the body. These cells are organized in units made up of army corps, just as every modern army is; only in the body such units are spoken of as "organs."

We can carry the analogy one step further. The movements of the several army corps, it is true, are controlled by the Headquarters Staff; but it must be evident that for complete coöperation, not only should the units be in touch with the General Staff, but also with one another. If the Staff breaks down, the army goes to pieces; if connections between the several units are broken off, the army also goes to pieces. Likewise with the body: a dis-

1

turbance in the brain immediately registers its effect on the rest of the body; so does a breaking-off of communications between the organs of the body.

Now while we have long been familiar with the functions of the brain, we have not until recently been aware of distinct means of intercommunication between the organs themselves, apart from their connection with the brain. The study of the ductless glands—the subject of this book—has led to this discovery, and with it the origin of a number of diseases has been made clear.

Glands.—When food is taken into the mouth it is met by a fluid called the saliva. Where does this fluid come from, and what is its function? This fluid, this saliva, is manufactured in factories situated in front of the ears. and between the lower jaw and the floor of the mouth. The factories get their raw materials from the blood, and the cells in these factories convert the raw materials into a product which we call saliva. Tubes connecting these factories with the interior of the mouth enable the manufactured fluid to be sent to the mouth. These tubes are spoken of as "ducts."

Now what is the function of this saliva, the product of cellular activity in these factories or "glands" of the body? It has in reality several functions, but there is one that stands out far above the others: it converts the starch of foods into chemically simpler products. This conversion is due to the presence in the saliva of "ptyalin," a substance that belongs to a class of compounds known as "enzymes" or "ferments." Much of the chemical work of the body—and this includes the plant as well as the animal kingdom—is due to the activity of these enzymes.

A gland, then, is an organ that has the power of taking certain materials from the blood and manufacturing from these raw materials a product which plays a part in the activities of the body. The salivary glands, wherein saliva is manufactured, are an example. The gastric glands lining the wall of the stomach, manufacturing gastric juice for the stomach needs, are another. The pancreatic and intestinal juices arising, respectively, from the pancreas and the small intestine, are still others. These examples could be multiplied.

Ductless or endocrine glands, or glands of internal secretion.—So far we have been considering glands that have tubes or "ducts" connecting them with an outer surface, such as the mouth or stomach. Glands are known, however, that have no such ducts, but that pass their products directly into the blood stream. Such glands are known as "ductless" glands. Sometimes they are spoken of as glands of "internal secretion," to distinguish them from those glands with ducts whose secretion is poured out on a surface. Sometimes they are called "endocrine" glands, "endocrine" being derived from two Greek words meaning "to separate within"; that is to say, an "internal secretion." [1]

It is these "ductless" glands that constitute the subject matter of this book; for recent research has shown us that they play an enormously important part in health and disease.

That the brain through its nervous mechanism controls the various parts of the body is common knowledge to-day; but that any portion of the body's activity should not be directly responsible to the brain for its controlling mechanism, is an idea that may sound revolutionary enough. Yet such is the case; and to illustrate it, a classical experiment due to Bayliss and Starling, two gifted English physiologists, will be described.

An illustration.—The food that we take into the mouth passes through the stomach into the small intestine. Here the food meets not only the intestinal fluid, but also the fluids coming from the bile on the one hand, and the pancreas on the other. The bile and the pancreatic fluid are led into the small intestine by means of tubes. Now why whenever food appears in the small intestine do bile and pancreatic juice also flow into it? The answer until recently was considered a very simple one. The

[1] This definition does not take into account the conception of the histologist as to what constitutes a "glandular" structure.

physiologists said that the nervous mechanism controls the flow of the fluids; that whenever food appears in the intestine, a nervous reaction calls forth the flow of bile and pancreatic juice. Bayliss and Starling showed that this conception needed modification.

These investigators, in experimenting with dogs, cut off all nervous connections with the small intestine; yet the fluids still continued to flow into it. They then suspected that possibly the acid from the stomach, upon reaching the small intestine, liberates something from the walls of the organ, which "something" finds its way to the pancreas and the bile, and thereby gives warning of the need of these fluids. They thereupon extracted a piece of the intestinal wall with hydrochloric acid—which is the acid found in the stomach—and injected this extract into the blood stream. There was an immediate and copious flow of pancreatic juice into the intestine.

What then happens in the course of digestion in the small intestine? The food that arrives from the stomach is acid, due to the hydrochloric acid that is formed in the stomach. This acid liberates a substance present in an otherwise inactive state in the wall of the intestine, and this substance travels through the blood to the pancreas, where it stimulates that organ to discharge its fluid.

Hormones.—Note that all this is performed with-

out any help from the brain.[1] Note that one organ
of the body—the intestine in our example—manufactures a substance which finds its way into the
blood stream and affects another organ of the
body, the pancreas. Here we have a classical example of the workings of a ductless gland; for in
every gland of the ductless variety a specific substance is manufactured that finds its way into the
blood stream and influences another organ or organs of the body. The substance so manufactured
is called a "hormone" (from the Greek "to excite"
or "arouse") or "chemical messenger." The hormone in the intestinal wall has been given the
name of "secretin" by its discoverers. Without
this secretin no pancreatic juice could find its way
into the intestine, and without pancreatic juice
no digestion of food could take place.

The small intestine is an example of a tissue
which gives rise both to an internal and an external secretion. Its internal secretion, the secretin, has already been described. But as a matter
of fact it also develops a secretion, the intestinal
juice, which is carried by ducts to the surface of
the intestine, in the same way that salivary or

[1] I do not want the impression to be conveyed that there is no
connection between the hormones on the one hand and the nervous
system on the other. As a matter of fact there are connecting
links, as the chapter on nervous disorders will attempt to make
clear. All I want to point out at this stage is that a particular
mechanism, for which the brain is held responsible, can be explained without involving the brain at all.

gastric juice is, and which, like these two, plays an important rôle in the digestion of foodstuffs.

The double property of producing both an internal and an external secretion, which characterizes the small intestine, is found in a number of other tissues. The external secretion of the pancreas, for example, is the pancreatic juice, which is carried by means of a duct to the small intestine. But it has been shown that the removal of the pancreas, or, what is the same thing, the removal with it of an internal secretion developed by the organ, and which finds its way into the blood, gives rise to the dreaded sugar disease commonly known as "diabetes." It would seem, therefore, that the utilization of sugar by the liver—the organ that stores sugar and gives it out when necessary—is controlled by an internal secretion developed by the pancreas.

The generative glands (ovary and testicle) show an internal and external secretion. The external secretion contributes to the reproduction of the species; the internal secretion plays a part in the development of male and female characteristics. One has but to think of the eunuchs in oriental countries, or of animals from which, for commercial reasons, the generative glands are removed, to realize what effects are produced by removal of these glands. But this will be treated more extensively later.

The thyroid, the pituitary, the sexual organs, the adrenals, all contain hormones, and much of the influence these organs exert upon our general well-being is due to these chemical messengers.

The composition of hormones.—Just what the hormones consist of is not clear in most cases. Yet the physiological chemist has been able to isolate in a pure condition a hormone present in the adrenal glands, and another in the thyroid; and the organic chemist has been able to start with very simple chemicals and synthesize these hormones in his laboratory. So that at least in two instances we know what the composition of a chemical messenger is; and the organic chemist will tell you that their composition is by no means terrifying.

Relation of hormones to vitamines, etc.—Ridiculously small amounts of hormone are sufficient to restore the normal equilibrium of the body. That is to say, where, as in thyroid deficiency, the disease can be cured by the administration of the corresponding hormone, the amount necessary is almost infinitesimal. Neither is the percentage of hormone in the active gland anything but slight. For example, Kendall, of the Mayo Clinic at Rochester, found that he had to use 6,550 pounds of fresh thyroid in order to get one *ounce* of thyroxin, the thyroid hormone!

These facts immediately suggest a relation between hormones and vitamines, and other substances present in the body in small amounts, but which nevertheless exert powerful effects—such as the enzymes or ferments.

Vitamines, as every reader of the press must know by this time, are unknown factors in food, probably present in amounts that defy weighing by the ordinary chemical balance. They are necessary for a continuation of the life cycle. As I have said elsewhere,[1] without vitamines there can be no life.

To what extent are we justified in comparing vitamines with hormones? As we have indicated, both are present in minute quantity, and a small quantity seems to go a long way. Both are therefore "catalytic" in their action; that is to say, they accelerate or hasten chemical action, without themselves undergoing any permanent change.

There are one or two direct clinical observations that are of interest also. Professor Dutcher, of the Pennsylvania State College, has performed a number of experiments which show that thyroxin, the thyroid hormone, has anti-neuritic properties; which means that it, like yeast, for example, can cure birds suffering from polyneuritis,—a disease first shown by Funk to be due to a lack of one of the vitamines.[2] If this is so, thyroxin has vitamine-like characteristics.

[1] See the author's book, *Vitamines: Essential Food Factors.* New York, E. P. Dutton & Company, 1921.
[2] See the chapter on Beriberi in the author's book on Vitamines.

Then again, McCarrison, an English investigator stationed in India, has shown that the adrenal
glands of birds suffering from polyneuritis are
much enlarged, and that the adrenaline content of
these glands is increased. As we shall see, the
adrenals, like the thyroid, are glands of internal
secretion, and adrenaline is the active hormone
present in these glands. Even more remarkable
is his discovery that though the adrenals become
enlarged, the other ductless glands degenerate and
tend to disappear altogether. It would seem as if
there were some connecting link between vitamine
B (the vitamine the absence of which causes polyneuritis) and adrenaline; as if a diet containing
adequate amounts of vitamine B had a restraining
influence upon the output of the adrenal hormone.

But we must not stretch these points too far,—
at least, not until these experiments have been repeated and extended. In the meantime it is well to
point out some obvious differences, if only to
strengthen our judicial attitude.

Vitamines are very susceptible to heat; or more
accurately, to a combination of heat and oxidation
(exposure to air) ; hormones do not seem to be
destroyed at a temperature even of boiling water;
at least, the last statement is true of secretin (the
intestinal hormone) and one or two others that
have been studied. Then again, in two instances—
adrenaline (from the adrenal gland) and thyroxin

(from the thyroid)—hormones have not only been isolated in the pure state, but they have actually been synthesized in the laboratory from relatively simple compounds. So far, not only have we been unable to synthesize a vitamine, but we have even been baffled in our attempts to isolate one in a pure condition.

It would, perhaps, be more in harmony with what we know to compare hormones with amino-acids, substances that are obtained when the proteins of food are broken up by the enzymes in our digestive system, or by acids used in the chemist's laboratory.[1] Chemically, thyroxin shows striking relations to tryptophane; and adrenaline, certain, though not such striking relations to tyrosine. Both tyrosine and tryptophane are among our best-known amino-acids.

One important point that is brought out in a discussion of this kind is to emphasize the importance of the "littlest things." Enzymes (or ferments), the substances that act on our food in the digestive tract, the compounds that are largely responsible for much of the metabolic activity of every cell, whether plant or animal, have long been known to illustrate the property that certain substances possess, of bringing about chemical changes in a large quantity of material, though the

[1] See the chapter on Amino-Acids in the author's book on Vitamines.

enzyme present may be there in relatively small amount. Still more remarkable are such reactions in that the enzymes do not seem to undergo any permanent change: at the end of the reaction we still find our enzymes, and for all the balance tells us, in the same amounts as before the reaction.

Substances that act like enzymes are called "catalysts." A catalyst may be defined as a substance that accelerates a chemical reaction without itself undergoing any permanent change. Hormones and vitamines are probably catalysts.

These catalysts are not confined to substances that take part in the activities of the living organism. The best method for the manufacture of oil of vitriol, or sulphuric acid, is by the employment of platinum as a catalytic agent: provided all impurities are rigidly excluded, the same platinum can be used over and over again. The manufacture of synthetic ammonia by the Haber process involves the use of a catalyst, probably nickel. I say "probably" because the details of the process are carefully guarded as trade secrets. At any rate, the production of ammonia from the nitrogen of the air, and the hydrogen that can be obtained either from the electrolysis of water, or as a by-product in the manufacture of lye, can be brought about with the help of a catalytic agent. Since this Haber process is on the road to solving the "nitrogen-fixation" problem, and hence many of our fertilizer difficulties, the catalyst in industry seems destined to play a part no less important than the catalyst that helps to maintain the chemical equilibrium of the body.

CHAPTER II

THE THYROID

This is the most frequently referred to of the ductless glands, not necessarily because it is the most important, but because much of the history of the ductless glands centers around this one; and also because of the success that has attended the treatment of at least one form of thyroid disease.

The thyroid, which usually weighs anywhere between one and two ounces, is situated in the neck. It consists of two parts on either side of the larynx (organ of voice) and windpipe, usually connected by a narrow strip of tissue. In contact with the thyroid are the "parathyroids," still smaller bodies, which were for a long time not sharply distinguished from the thyroid proper, and which as a consequence gave rise to much confusion in the interpretation of experimental results. These parathyroids will receive treatment presently.

That the thyroid is an organ that plays an all-important part in the various activities of the body becomes apparent when for any reason it behaves

14

abnormally. The fluid, or still better, the active hormone that the body secretes, is under certain conditions largely increased. We get then a condition of hyper-thyroidism. The disease known as "exophthalmic goiter" (Graves's disease and Basedow's disease are synonyms) is probably a case in point. On the other hand, a condition may arise wherein the quantity of secretion and the supply of hormone become deficient. This may develop "myxedema" in the adult and "cretinism" in the infant; such being examples of hypo-thyroidism. We shall take these up in turn, and we shall begin with hypo-thyroidism.

Hypothyroidism.—In medicine definite clues as to the type of disease are usually obtained by preliminary experiments with animals. We find that as early as 1859 Schiff, a Swiss physiologist, investigated the effect of thyroidectomy on animals ("thyroidectomy" is a convenient word to describe the idea of the surgical removal of the thyroid. "Dectomy" is derived from a Greek word meaning "excision.") His results, though suggestive, were inconclusive, due to the fact, as we know to-day, that Schiff removed the parathyroids as well as the thyroids. But this physiologist went a step further. Having removed the thyroid, he next investigated the effect of implanting thyroid from an animal of the same species. The results in a number of cases were highly encouraging.

Fired by the work of the English physicians Gull (1872) and Ord (1878), and the German Kocher (1883), Schiff, in 1884, published his celebrated paper, "On the Effects of the Removal of the Thyroid Body," in one of the Swiss medical journals. How clearly Schiff recognized the gland to be an internal secretory one may be gathered from this quotation: "We may wonder if the thyroid body produces in its interior a substance which it delivers into the blood stream and which constitutes a nutritive element for another organ (nervous), or whether it acts mechanically by its anatomical position. To decide between these two alternatives, it is necessary to find a means of transplanting it, by grafting it into another part of the body. If, after this has been done, the accidents resulting from its removal are avoided or reduced to a minimum, it is evident that the action of the thyroid is due to its composition and not to its anatomical relations; this will prove the thyroid to have a chemical function."

The grafts Schiff tried disappeared in time, but not before it was noticeable that there was an improvement in the condition of the animal. From these results he concluded that "the substance of the grafted organ, taken up by the blood, serves to counterbalance the untoward effect of thyroidectomony";—clearly a case of hormonic action.

He had another idea which, however, he did not

put to the test of experiment, and thereby fell short of another great discovery: "It would be curious," he writes, "to investigate if the macerated extract of the thyroid, introduced into a cavity, or injected into the rectum, has the same immunizing power."

A repetition of Schiff's experiments by other investigators did not always corroborate Schiff's findings. We have already suggested that an incomplete knowledge of the parathyroids may have been the cause of this. We may also add that not only do animals of different species behave differently, but even animals of the same species do, provided there is a marked difference in age.

Symptoms.—Where the operation in an animal has been successfully performed, some such symptoms as the following develop: The skin becomes thick and dry; there is a loss of hair; the animal shows a tendency towards obesity, particularly in certain portions of the body; the muscles become weak and the tissues renew themselves but slowly. The blood is poor, in quality and quantity; the temperature is below normal; and the sexual functions are interfered with. This interference with the sexual glands is of uncommon interest, since it suggests a close interrelationship between the various hormones of the body—an observation that is strengthened every time a disease due to a hormonic disturbance is examined. The nervous system is also attacked, dullness and general apathy being markedly apparent. A histological examination shows many of the nerve cells to have shrunk in size.

Cretinism.—If instead of removing the gland it atrophies or wastes away, the symptoms are quite identical. In a child where such a condition occurs we find growth to be arrested. The head and face look deformed; the expression is decidedly idiotic. The face is pale, the hair thin, the skin dry, the abdomen swollen. The development of the generative organs is delayed. Deaf-mutism is quite common. These are all symptoms of the "cretinous" child, the disease being spoken of as "cretinism."

Professor Falta describes a cretinous child, four and one-half years old, as follows: "Head at birth already large. Speech up to second year of life consisted of the simplest words only, such as 'tata,' 'mama'; and since this time the child has not spoken much otherwise. Head extremely large. Very low forehead, eyes stand wide apart. Saddle-nose; thick broad tongue that protrudes from the mouth. Cheeks very thick, throat very thick and stubby. Thyroid not palpable. Thick hair on back. Skin of the body springy, elastic; hands and fingers chubby. Abdomen much distended. The child often stares into space for a long time, but at times is lively and cries loudly. No trace of speech. Puts all objects into his mouth. Impressions of hearing entirely absent; no reactions of the eyelids to sounds."

If now there still remains some question as to the soundness of the diagnosis, all doubts are immediately removed by prescribing extracts of the thyroid gland to the young sufferer. The recovery as a consequence is little short of miraculous.

Thyroid feeding and the development of frogs.—In his fascinating article on "Natural Death and the Duration of Life" (*Scientific Monthly*, December, 1919), Professor Jacques Loeb describes the remarkable effect of thyroid feeding on the development of the frog and salamander. He writes: "It is possible that some of the changes underlying metamorphosis are due to changes in the circulation of the blood. Gudernatsch made the remarkable discovery that this metamorphosis, which in our climate usually occurs during the third or fourth month of the life of the tadpole, can be brought about at will even in the youngest tadpoles by feeding them with thyroid gland, no matter from which animal. By feeding very young tadpoles with this substance, frogs not larger than a fly could be produced. Allen added the observation that if a young tadpole is deprived of its thyroid gland, it is unable ever to become a frog; and that it remains a tadpole which can reach, however, a long life and continue to grow beyond the usual size of a tadpole. When, however, such superannuated tadpoles are fed with thyroid they promptly undergo metamorphosis.

"These observations cleared up an old biological puzzle. Salamanders also undergo a metamorphosis which is, however, less striking than that of the tadpole of a frog. In the salamander the metamorphosis consists chiefly in the throwing off of the gills, and in changes in skin and tail. In Mexico a salamander occurs which through its whole life maintains its tadpole form, namely, the axolotl. Attempts to induce the axolotl to metamorphose failed until after Gudernatsch's discovery an investigator fed the axolotl thyroid gland, and this brought about metamorphosis. . . . It seemed possible that the iodine contained in the thyroid was the active principle causing metamorphosis in tadpoles. This was confirmed by Swingle who succeeded in inducing metamorphosis in tadpoles by feeding them with traces of inorganic iodine."

Administration of thyroid extract.—Thyroid extract may be administered in one of three ways: by mouth, by injection beneath the skin (subcutaneous), or by injection into a vein (intravenous). The last method yields the quickest response. This might be expected if we remember that the thyroid hormone, and all other hormones, reach the various parts of the body by means of the blood stream.

Usually, however, the physician prefers to mix the thyroid extract with the patient's food, this being a much simpler proceeding, and in the long run just as effective.

Thyroid gland transplantation.—Why when the administration of thyroid gland is so efficacious, thyroid gland transplantations should be undertaken to cure patients suffering from hypothyroidism, is not clear. I take up a New York newspaper for Dec. 9, 1920, and in it is the following (special dispatch from Chicago): "Mary Zembok, 19, of Joliet, Ill., may become a normal girl of 19—that is if the grafting of a monkey gland into her neck to-day restores her mentality and physical development. Physicians who performed the operation said that it had been successful. Mary had been mentally defective almost since birth, and comparatively paralyzed. The mother, a Pole, has five children besides Mary, all of them normal. When Mary was two years old it became apparent that she would not develop into a normal child, and Mrs. Zembok relegated her to the basement of their home, where she lived in perpetual darkness until two months ago, when the health authorities learnt of the case. The child, when found, was immediately removed to a hospital here. It was decided that the only hope for a full recovery would be the gland transplantation operation. To-day the monkey, a full-grown animal, and the child were taken to the operating room, the thyroid gland removed from the animal and transplanted into the neck of the child."

I submit that this is excellent material for a novel, play or "movie."

As I write this paragraph I am informed that J. R. Brinkley, M.D., C.M., Sc.D., Chief Surgeon, Brinkley-Jones Hospital and Training School for Nurses, Milford, Kansas, graduate of the medical department of Loyola University, who has travelled "all over the world"—that this same doctor has completed a 96-page book on the Goat Gland Transplantation. Dr. Van Buren, reviewing the book for the *New York Times,* says: "A fair commentary on this book, I should say, is that one of its outstanding features is its delightful naïveté."

Is the absence of hormone responsible for cretinism?—The very fact that administration of thyroid extract causes recovery in the case of cretinism, a specific disease, points to the presence of some hormone in the thyroid that is responsible—some substance, then, that reaches the other organs of the body by means of the blood stream. Unfortunately for the welfare of the community, the other ductless glands do not exhibit this specificity to the same marked degree. With them the situation that arises is often of a character so complicated as to baffle the sharpest intellects of the medical profession.

Myxedema.—The atrophy of the thyroid in the adult gives rise to a condition quite similar to that described in the child. Here, however, instead of calling the disease "cretinism" we call it "myxedema," to denote the mucous fluid that gathers beneath the tissues and that gives rise to swellings all over the body. There is a belief that the immediate cause of this disease is due to the presence of an excess of "mucin," a substance that gives the "ropy" consistency to saliva, and that is also found in various parts of the body. The thyroid hormone presumably prevents the accumulation of an excess of mucin.

Symptoms in myxedema.—A person afflicted with myxedema, which, by the way, is more common in woman than in man in the proportion of at least two to one, gradually assumes an appearance that is beyond all recognition. The dull mind, the sluggish movement, the unsteady gait, combine with the general alteration of features to make the person a most pitiable spectacle. The patient may eat quite little, yet so poor is the assimilatory mechanism, that even that little is not easily taken care of; so that there is a marked accumulation of food reserve in the body, and the individual becomes abnormally fat.

Treatment with thyroid extract.—Here again, as with the child, cures may be obtained by the administration of thyroid extract. But in both cases the extract must be administered at regular intervals and kept up indefinitely, otherwise there is a relapse. The discordant results that have been reported with thyroid feeding in myxedema are almost invariably due to the type of thyroid extract used. The various samples of these on the market vary greatly in efficacy, due no doubt to the method and source of extraction. Two commercial houses justly famous for the preparation of such extracts are Burroughs Wellcome, of London, and Parke, Davis, of Detroit, Mich. Now that Kendall has succeeded in isolating the active principle of the thyroid, there need be little occasion for future variation.

Evidently the administration of the thyroid substance does not give rise to any accumulation of the material in the system. A certain amount is, as it were, used up in each operation. Under normal conditions, when the gland in the body functions properly, the amount of hormone necessary for regulating metabolic processes and nerve responses is manufactured whenever required. This, of course, is no longer possible when the gland is in an atrophied condition.

Many cases are on record which show the development of myxedema when the thyroid is removed—removed because, let us say, of a tumor growth. Both the removal of the thyroid and the atrophy of this organ give rise to the same disease.

A number of investigators—among them the justly famous Mayo brothers of Rochester—showed that where surgical removal of the thyroid becomes imperative, myxedema may be prevented by leaving intact at least one-fourth of the organ.

Iodine in the thyroid.—It would have been strange if no attempts had been made to isolate the hormone or active principle present in the thyroid gland. As a matter of fact, many such attempts were made, particularly after the successful isolation of the active principle of the adrenal glands. The first important discovery that was made in this direction was by Baumann in 1895, who showed that the thyroid is rich in the element iodine. Until Baumann's time no one had had the slightest suspicion that such an element as iodine existed in the body; yet a careful analysis of the thyroid showed it to be there in appreciable quantity. That naturally suggested two things: that the active principle of the gland was an iodine compound, and that iodine, or more likely foods containing iodine, have to be supplied to the body.

Isolation of the thyroid hormone.—Much work has since been done on the chemistry of the iodine compound in the gland. Various substances have been isolated from it containing percentages of iodine greater than that in the whole gland, showing that the preparations, if not pure, did at least represent concentrated fractions. These substances were given various names, and their discoverers claimed that their iodine compounds could do all that the gland extract itself could do: that, in short, such substances could be employed in the place of the gland extract in the treatment of cretinism and myxedema. Many have been the statements for and against the use of such preparations. Lately, however, E. C. Kendall, working at the newly-created Mayo Foundation in Rochester, has, after some ten years of intensive work, actually succeeded in isolating the iodine compound in a pure form. He calls it "thyroxin," and the compound contains no less than 60 per cent. of iodine. It has already been extensively used by Kendall and others in the treatment of thyroid deficiency diseases, with marked success in almost all cases. "This could be made even stronger, as we have not found a single case of thyroid deficiency that has not responded to an intravenous injection of thyroxin; and, furthermore, the response is a quantitative one. That is, for every milligram (one-thousandth of one gram, and 2½ grams correspond to one ounce) injected, the basal metabolic rate (see p. 39) increases 2½ per cent." (E. C. Kendall.) This constitutes another triumph for the chemist in his application of chemistry to medicine.[1]

---

[1] *Kendall's work.* This work of Kendall's is of such importance that it warrants further discussion. As, however, some knowl-

Endemic goiter.—Before dismissing the subject of hyposecretion a word must be said about the cases known as endemic goiter. In such cases the front part of the neck becomes swollen, due to the enlargement of the thyroid gland (hence the name

edge of chemistry must be assumed at this point, I would suggest that those readers who do not possess such knowledge had better skip this footnote.

The first detailed account of the isolation of the active principle appeared in 1919, when Kendall published his paper, "On the Isolation of the Iodine Compound Which Occurs in the Thyroid," in the *Journal of Biological Chemistry.* The research was begun in 1910, so that no less than nine years of continuous work were spent in attempts to isolate the hormone. Up to 1919, 33 grams (a little over an ounce) of the hormone "thyroxin" had been isolated from 6.550 *pounds* of fresh thyroid.

Kendall's method of isolating thyroxin is as follows: Fresh thyroid gland is hydrolized with sodium hydroxide. The fats are removed by rendering the sodium soaps insoluble, and the clear alkaline filtrate is acidified. Acid soluble and acid insoluble portions are obtained, each containing one-half of the total iodine. (Kendall has confined his attention to the acid-insoluble portion, which contains all the thyroxin, but what type of iodine compound is in the acid soluble portion, and to what extent that iodine combination is of importance, remains to be seen.) The acid insoluble portion is filtered off, the precipitate redissolved in sodium hydroxide and reprecipitated with hydrochloric acid. The substance is next air-dried and dissolved in 95 per cent. alcohol. A hot, concentrated, aqueous solution of barium hydroxide is added to the alcoholic filtrate, and the mixture heated under a reflux condenser. This is next filtered, a small amount of sodium hydroxide added to the filtrate, and carbon dioxide passed through the solution. Barium and sodium carbonates are removed by filtration, and the alcohol is removed by distillation.

The product is purified by redissolving in alcoholic sodium hydroxide and again passing in carbon dioxide. The sodium carbonate is filtered, and the alcohol evaporated. The last traces of alcohol are removed by heating on a water bath. At this point the monosodium salt of thyroxin separates. The thyroxin itself is obtained by dissolving the salt in alcoholic alkali, and precipitating with acetic acid.

"goiter"), and a pressure is exerted on neighboring glands such as the trachea and esophagus. The
enlargement of the thyroid gland would lead one
to suspect that there is a hyper-, rather than a hypo-
secretion of the hormone. This, however, is not
true of this type of goiter, which is quite prevalent
in Switzerland, and in the Lake sections of our
country (hence called "endemic" to denote its local
character). That we here deal with a type of hypo-
secretion is made evident by the general symptoms

Thyroxin is 4,5,6 tri-hydro-4,5,6 tri-iodo-2-oxy-beta indolepro-
pionic acid

It is a colorless, odorless, crystalline substance, insoluble in
aqueous solutions of all acids, including carbonic; soluble in
sodium, potassium and ammonium hydroxides; slightly soluble
in sodium and potassium carbonates. It forms salts with metals
and acids. It contains 65 per cent. of iodine.

The above formula was established by an analysis of the substance, and by a study of its derivatives.

An interesting observation, made by Dr. Dutcher, but not yet
confirmed, so far as I am aware, is to the effect that thyroxin
has vitamine properties; it behaves like yeast, for example, in
curing birds of polyneuritis and men of beriberi (see the chapter
on Beriberi in the author's book on Vitamines). Are we after
all on the eve of discovering some relationship between vitamines
and hormones?

that develop, and by the ready response to treatment with thyroid extract. Poultices of burnt seaweed—a fruitful source of iodine—and iodine itself painted on the skin, were used for this disease long before it was known that the thyroid contained this element.

Iodine an essential element.—Now as to the iodine itself, is it really essential to the body? An element can neither be synthesized in the laboratory nor in the body; it can merely be obtained from nature or from compounds that already contain it. If we deprive our food of all iodine we deprive the thyroid of what seems to be its essential element. Yet the situation is not as clear as it seems; for though many experiments have shown that thyroid activity and the iodine content of the gland go hand in hand, yet cases are known—of animals, it is true—where the thyroid contains no iodine at all, and yet the animal seems quite normal. On the other hand, in endemic goiter some remarkable cures have been obtained by the administration of simple inorganic iodides. Marine and Kimball, of the Western Reserve University, Cleveland, started in 1917 with a survey of the incidents and types of thyroid enlargement in the schoolgirls of Akron, Ohio, from the fifth to the twelfth grades inclusive. For treatment they used sodium iodide, which was taken in doses of three grains daily for ten consecutive school days, repeated each spring and autumn.

Recently (June, 1920) they published results of studies which covered a period of some thirty months. Of 2,190 pupils that had received the sodium iodide treatment, five had shown enlargement of the thyroid, while of 2,305 pupils who had not received any such treatment, 495 showed an enlarged thyroid.

In a later article (Oct. 1, 1921) they write: "Of 1,182 pupils with thyroid enlargement at the first examination who took the prophylactic, 773 thyroids have decreased in size; while of 1,048 pupils with thyroid enlargement at the first examination who did not take the prophylactic, 145 have decreased in size. . . Klinger has recently (1921) reported even more striking curative results in the school children of the Zürich district. He worked with school populations in which the incidents of goiter varied from 82 to 95 per cent., while our maximum incidence in Akron was 56 per cent. With such a high natural incidence of goiter, his observations naturally deal more with the curative effects. Thus of 760 children, 90 per cent. were goitrous at the first examination. After fifteen months' treatment with iodine, only 28.3 per cent. were goitrous, of a total of 643 children re-examined."

Hypersecretion.—An excessive secretion (hypersecretion) developed by the thyroid gland may give rise to the disease commonly known as "exophthalmic goiter," though the English sometimes call it "Graves's disease," and the Germans, "Basedow'sche Krankheit," to denote the work done by pioneers. Often these diseases present a greater complexity than a mere hypersecretion. Here again much of the preliminary work was done by experiments on animals. Instead, however, of removing the thyroid gland—which we do when we study the effects of hyposecretion—extra doses of extracts of the gland are administered. When this is done we almost invariably produce symptoms in the animal that bear a strong resemblance to exophthalmic goiter in man.

Symptoms in exophthalmic goiter.—As might be suspected, the symptoms of this disease are very much the reverse of those in cretinism and myxedema. The slowing up of cellular processes, so characteristic of myxedema sufferers, gives place to a decided acceleration of these processes. The stupid, apathetic expression is replaced by an anxious, restless one. In the place of a deposit of fat there is a wasting away of the tissues. The subject becomes thin because of excessive metabolism (we shall take up the question of metabolism later). The pulse is rapid—it may vary from 100 to 140 per minute—and irregular. The thyroid is usually, though not invariably, increased in size (hence "goiter"). There is also usually, though not invariably, an abnormal protrusion of the eyeball

(hence the name "exophthalmos,") giving the subject a "startled" expression; and this goes hand in
hand with his anxious, restless appearance. Very
often the patient suffers from profuse perspiration.

The Romans were evidently not unfamiliar with
this disease, for it is stated that they refused to
buy a slave if he had an enlarged goiter, or if he
showed protruding eyeballs, on the very practical
grounds that he was not a fit subject for hard
work.

One or more of the symptoms enumerated show
themselves in the individual suffering from exophthalmic goiter; but one symptom that is always
present and that, so to speak, gives the clue to the
type of disease, is the excessive rapidity in the
action of the heart (tachycardia).

Not distantly related to the physical condition
of the patient is his mental state. The relation of
the ductless glands to the general mental make-up
of the individual is of sufficient importance to warrant special treatment, and a chapter will be devoted to that subject later on. Here it may be
mentioned that the patient suffering from exophthalmic goiter becomes, as Dr. Cobb puts it, an
"intractable, selfish, restless and inconsiderate
being. The medical attendant as a rule receives
the full benefit of this, and can do little that is
right. He is either old-fashioned when he explains
that the reason for rest in bed, for example, is to

avoid straining an already weakened heart; or an ignoramus if he insists that the rest combined with hygienic principles offers the best hope for alleviating the disease. If he suggests trying a new remedy he is experimenting with her; if he persists with the old he is a 'stick-in-the-mud.' "

Dr. Cobb in his example deliberately uses the feminine gender because the disease is really far more common in the female than in the male. The proportion is placed by some authorities as high as five to one. An attempted explanation for this "partiality" is based on the fact that in the female the gland is enlarged usually at puberty and during pregnancy.

Treatment.—When we come to the means at our disposal in bringing about a cure, we find no such definite road to success as in cases of hypothyroidism. There, as will be remembered, definite amounts of extracts of the thyroid gland, when regularly administered, changed the secretion of the gland from subnormal to that approaching normal. Here, with an excessive secretion, it might be surmised that a partial removal of the gland would be successful in restoring the health of the sufferer. This has been tried many times, and with success in a fair number of cases. But the operation is a difficult one, and only a surgeon of extraordinary skill and much experience can perform it; and even then a successful operation is often followed by post-operative changes that are quite discouraging—a factor that of course enters into many other types of operations. In any case, unless the disease has taken a very grave turn, when the individual is pretty much on the borderland between life and death, the treatment is apt to be anything but surgical.

Treatment other than surgical centers around the word "rest." No particular insight on the part of the lay reader is required to understand the reason for this. If the body machine is performing its cycle of operations at an accelerated speed, anything that will tend to decrease the rate will prove beneficial. Dr. Charles Mayo says: "The opinion of an eminent surgeon (Kocher) that 90 per cent. of all goiters can be improved so as to make operations unnecessary, was probably based upon observations of the effect of rest, for rest is the common element in all the various forms of treatment that have proved successful."

The attending physician has, of course, to consider the question of diet and the use of drugs. Food containing iodine should, *ipso facto,* be barred from the table, on the assumption that since the hormone responsible for thyroid activity is probably an iodine compound, we need do nothing to increase the quantity of iodine in the body. Then again a drug that will tend to decrease the rapid action of the heart may prove advisable. These items can be safely left to the discretion of the medical adviser.

So far attempts to find some direct and simple cure for exophthalmic goiter have failed. The "rest" cure in many cases, and removal of part of the gland in a number of cases, have proved successful. Other methods of "cure" include the production of a substance that will act as an antitoxin, and neutralize the excessive quantity of hormone present; the application of X-rays; and the administration of extracts of parathyroids or the thymus gland, or of calcium salts.

It may not be amiss to point out at this time that the psychic factor in the treatment of hyperthyroidism cannot be overlooked. Psycho-analysis, handled by pseudo-scientists, has become a laughing stock, just as glandular treatment and the general subject of the ductless glands threatens to become; but psycho-therapy, practised by the skillful physician, is at times of inestimable aid in putting the patient on his feet.

Metabolism studies.—Cases of hyper- and hypothyroidism have fairly well-recognizable symptoms. These have already been discussed. But we have to describe a method which helps the diagnosis a great deal. This depends upon the fact that the thyroid is *par excellence* the organ that regulates the metabolism of the body.

But before we go any further we must explain

what we mean by the word "metabolism." Food is taken into the system, and with the help of enzymes or ferments and the oxygen of the air, undergoes chemical changes in the body, yielding carbon dioxide, moisture, and various nitrogenous products that are eliminated principally through the kidneys. The processes involved are digestion, assimilation and excretion. Under the heading of metabolism we may include all those changes that occur in foodstuffs from the time they are absorbed to the time they are excreted. Huxley used the word "metabolism" to "denote the sum total of those chemical changes which take place in living matter, and in virtue of which we speak of it as living" (metabolism comes from the Greek meaning "change.")

Lavoisier.—Lavoisier, a Frenchman, more than a hundred years ago showed that valuable information as to the metabolic changes that go on in the body can be obtained in one of three ways: by estimating the amount of oxygen consumed, or of carbon dioxide eliminated, or of heat evolved. With regard to the heat evolved, it must be remembered that the reactions of the body, like all chemical reactions, are accompanied by temperature changes—usually by an increase of temperature, as in the case of the body.

Since Lavoisier's time, chemists and physiologists

have given much attention to the study of "respiratory exchange"—to the relationship of oxygen intake and carbon dioxide output,—and to the measurement of the heat evolved in the reaction. Various types of calorimeters have been invented for that purpose.[1] In Germany, Voit, Pettenkofer, Rubner and Zuntz, and in this country, Atwater, Rosa and Benedict, have done much to advance our knowledge in this direction. The sum of such knowledge has been to supply us with exact data regarding the heat evolved in individuals under varying conditions. If a certain amount of heat is evolved, that amount of heat must also be supplied, and the supply of such heat can come only from the food supplied and "burnt" in the body. That is how we arrive at certain fundamental food requirements.

Constant Temperature.—Not the least remarkable of the many remarkable phenomena noticeable when we study the living organism is the way the temperature within us remains constant. The temperature outside may vary considerably, yet under normal conditions the temperature within us varies very little. A constant temperature means a constant heat production.

Of course this is true only of warm-blooded animals, not of cold-blooded ones, such as the frog;

[1] See the chapter on Calories in the author's book on Vitamines.

its temperature is but slightly higher than that of its environment, and its metabolism, therefore, varies considerably from winter to summer.

Surface area.—Though for a long time the metabolism (measured in calories) of man and warm-blooded animals was taken to be proportional to their weight, Rubner, of the University of Berlin, has shown that a closer relationship is obtained if instead of the weight we substitute the surface area of the man or animal—that is to say, the area of the surface exposed. Then we get such figures as the following:

	Weight in kilograms (1 kilo equals 2¼ pounds)	Calories produced Per kilo	Calories produced Per square meter of surface
Horse	441	11.3	948
Pig	128	19.1	1078
Man	64.3	32.8	1042
Dog	15.2	51.5	1039
Goose	3.5	66.7	969
Mouse	0.018	212.0	1188

(A calorie is the amount of heat necessary to raise one kilo of water one degree centigrade. A meter is a little over a yard.)

Making allowances for experimental difficulties and inaccuracies, the last column of figures is fairly constant, unlike the second, which varies considerably. We may say then that there is an evenness of heat production per unit of body surface.

The difficulty in referring to surface area is a

difficulty connected with measuring such an area. Mathematicians have not been wanting to devise a formula which connects weight with surface area. All you have to do is to weigh your man and then multiply the cube root of the body weight squared by 12.3, and you have your surface area! Du Bois, however, working at the Russell Sage Institute of Pathology, connected with the Bellevue Hospital, New York, has shown a method by which the surface area can actually be measured. "He covered the body surface with light-fitting underwear, applied melted paraffin, and then paper strips to prevent change in area when the covering was removed. This model of the surface when cut into flat pieces was photographed upon paper in which equal areas were of equal weight. From the weight of paper which received the photographic impression the area of body surface could be calculated." [1]

So accurate is this method, that a ball having an area of 0.1490 square meter, when measured in this way, gave the figure 0.1488. Du Bois also showed that the calculation of surface area by the formula method involved an average inaccuracy of 16 per cent.

Basal Metabolism.—Using the Du Bois method, the heat production per square meter of surface is 39.7 calories per hour, provided the individual is resting and is "normal," and provided the experi-

[1] Lusk, *The Science of Nutrition*, p. 214.

ment is carried out before the administration of food in the morning (that is, after he has fasted for 12 to 14 hours). This is the so-called "basal metabolism," and constitutes the unit of reference whereby we can measure what are the deviations from the normal.

Metabolism in thyroid disease.—All this is a long, but, I believe, a necessary preface to what we are now coming to: metabolism in thyroid disease. We started out by saying that the thyroid, by means of its hormone, regulates, perhaps more than any other organ of the body, the amount of metabolism in the body. Under normal conditions, with a normal thyroid, the metabolism of the individual is such as to show a normal "basal metabolism." When, however, hyper-thyroidism sets in, with a gland that is excessively active, the rise in metabolism above the normal amount is considerable. *Vice versa,* where the patient suffers from hypothyroidism, with a decreased activity of the thyroid, there is a decided drop in the metabolic rate. For example, when the metabolism of a patient suffering from exophthalmic goiter (hyper-thyroidism) was investigated it was shown to be 75 per cent. *above* the normal basal of 39.7 calories (usually written +75); on the other hand, with a cretin the metabolism was 22 per cent. below normal (—22). Incidentally the same cretin, after being fed with thyroid extract, showed a normal basal metabolism.

Value of basal metabolic studies.—Of what value are these metabolic studies? Of the very highest in thyroid disease; for, taken together with the other symptoms, a metabolism experiment will confirm or disprove the diagnosis. But there is a still more important point to be brought out: once the diagnosis has been made and a method of treatment adopted, its progress can be followed by periodically carrying out such metabolic experiments. If the patient suffers from hyper-thyroidism, a successful treatment will gradually show a lowering of the metabolic rate; if from hypothyroidism, successful treatment will reveal an increase in the metabolic rate.

There is one drawback to the method: It is not easy to carry out, and requires skillful scientists to handle the instrument.

It may be mentioned in passing that metabolic studies have been undertaken in other ductless glandular diseases, notably the pituitary, but so far with little result; the plus and minus variations are not of a sufficiently pronounced character.

In any case we may conclude with Dr. Boothby, of the Rochester Clinic, that "the basal metabolic rate, which is a determination of the heat production in a person under standard conditions, serves as a measurement of the most fundamental process of life itself. Variations of the heat production, and alterations in the body temperature are to be considered rightly a means of fundamental disease classification, and the basal metabolic rate serves as an accurate diagnostic aid in the recognition of the presence or absence of hyper- (and hypo-) thyroidism."

All this is very true provided the diagnosis does not merely confine itself to a determination of the basal metabolism of the patient; for increases in basal metabolism are obtained in pernicious anemia or leukemia (increase in the number of leucocytes in the blood) or fever. These, however, can be readily differentiated from goiter afflictions.

CHAPTER III

THE PARATHYROIDS

Attached to the thyroid are four small organs, weighing in all not more than two grains, which are known as the "parathyroids" ("para" means "near"). For a long time these organs were not distinguished from the thyroid proper, and even when discovered and described, they were—and by some still are—regarded as part of the larger gland. The concensus of opinion, however, seems to be in favor of the belief that the parathyroids are independent organs with an independent action all their own; and that if they are related to the thyroid, the relationship is no closer than that of any other two ductless glands in the body. For much of our knowledge of the parathyroids we are indebted to the French physiologist, Gley.

Tetany.—The removal of the parathyroids in an animal, or their decay or removal in man, gives rise to the condition known as "tetany" (which, by the way, must not be confused with tetanus or lockjaw). This "tetany" is characterized by the

43

sudden involuntary contractions of the muscles and extremities. "The fits are sometimes frequent, but more often occur at long intervals." The respiratory tract, the heart, and the body temperature are also affected. The illness is very often prolonged, and often terminates fatally. There is a loss in weight, indicating a disturbed metabolism. The presumption that the metabolism of bone, and particularly that relating to calcium, is affected, has experimental support, in that the administration of calcium sometimes relieves the sufferer.

That tetany is the result of the absence of the parathyroids is clear enough when we consider that the administration of parathyroid extract relieves the patient, at least temporarily. You see, we are here dealing with a case of hypo-parathyroidism, just as in cretinism and in myxedema we dealt with instances of hypo-thyroidism; and just as the thyroid secretes a specific hormone the absence of which gives rise to cretinism in the child and myxedema in the adult, so the parathyroid secretes a hormone the absence of which gives rise to tetany. Notice that the most important piece of evidence for this view is that in both cases the symptoms disappear upon the administration of the extract of the appropriate gland.[1]

[1] The results with parathyroid extracts are more conflicting than those with thyroid ones. A safer position to take would be to say that the symptoms of tetany can be removed by "grafting" the gland, rather than by administering an extract of the gland.

Koch has found that in tetany there is an excessive elimination of guanidine, an important nitrogenous substance that plays a part in the metabolism of the body; and he has postulated the theory that tetany is due to an intoxication of the central nervous system by this guanidine,—a view that has a number of opponents.

It may now be asked, if hyper-secretion of the thyroid may give rise to exophthalmic goiter, what does hyper-secretion of the parathyroid give rise to? We are not very sure. We have reached the end of the land of fact and have now begun our journey in that of speculation. Since one of the chief objects in writing this book is to distinguish fact from fancy, and therefore to combat much that has been written on the subject of the ductless glands, we shall avoid entering this land of speculation.

The study of the parathyroids makes clear why in operations involving the removal of thyroid tumors, the earlier surgeons were much troubled, because very often an apparently successful operation would be marred by the development of tetany in the patient. We now can attribute such a result to the removal of the parathyroids as well as to that of the thyroid. The surgeon of the present day, engaged in an operation of this kind, invariably leaves at least two of the four parathyroids, for not the least remarkable of the many remark-

able facts gathered about these ductless glands is
that even when only part of the organ remains in
the body, it still functions, though, of course, not as
well as the entire organ would.

CHAPTER IV

THE PITUITARY GLAND

It seems hardly credible that a piece of tissue weighing one-sixtieth of an ounce, "the size of a hazel-nut," and lying at the base of the skull, should, perhaps, be involved in the production of giants on the one hand, and dwarfs on the other; that it, like the thyroid gland, should profoundly influence metabolic and brain functions, growth and life itself; yet such seems to be the case with the pituitary gland (sometimes called "hypophysis cerebri"). The conflicting results that encumber much of the work on this gland—and on other ductless glands for that matter—are due to the almost insurmountable difficulties that the surgeon encounters in locating and removing the gland, without at the same time bringing about secondary reactions that are due to causes other than the removal of the pituitary. Much of what we know we owe to the genius of Harvey Cushing, professor at Harvard.

The pituitary described.—The pituitary body, like the thyroid, consists of two parts, two "lobes," and is situated in a depression of a wedge-shaped bone lying at the base of the skull. It was an organ not unknown even to such ancients as Galen and Vesalius, who thought that it was involved in the formation of nasal secretions (hence the name "pituitary" from the Latin "pituita," meaning "phlegm"). Yet, as Professor Cushing has pointed out, Lower, as far back as 1672, in a paper entitled "Dissertatio de Origine Catarrhi" says, "For whatever serum is separated in the ventricles of the brain and tissues, out of them through the infundibulum to the glandula pituitaria distils not upon the palate, but is poured again into the blood and is mixed with it"—which is a very modern way of defining a ductless gland.

The vagueness attaching to the function of the pituitary remained such that the French were fond of calling it "l'organe énigmatique." A distinct advance was made in 1886 when Marie, a French scientist, associated a peculiar disease characterized by the enlargement of certain bones of the body, and which he called "acromegaly," with a tumor of the pituitary. The next step was a perfectly natural one; the tumor had to be removed. When this was done other strange symptoms made themselves manifest, due in most part, as we know to-day, to glandular insufficiency, and also, to some extent, to imperfect operations. Marie formed the opinion that "acromegaly" was a disease due to a deficiency of pituitary activity, and for years scientists blindly believed in this theory, attempting by extirpation experiments on animals to reproduce acromegalic symptoms. All such experiments resulted in failure. To-day we know the reason why. The acromegalic in all probability suffers from an excess rather than from a deficiency of pituitary hormone, or hormones.

Only since 1895 has intelligent guesswork given place to a fair amount of knowledge.

Removal of the pituitary.—Complete removal of the pituitary gland causes death. Innumerable experiments with puppies show that death results in from two to thirty days. Where the life of the animal is prolonged, a post-mortem examination reveals that a portion of the gland still remains. Cushing, as the result of some 200 such extirpations on dogs, has come to the following important conclusions: that a prolongation of life, if not a complete recovery, is possible by transplanting the gland into an animal that had had its pituitary removed; that whenever the animal survives the operation, it is because the gland, and more particularly the anterior [1] portion of it, has not been completely removed; and that when the posterior portion alone is removed, the animal does not die. It follows then that the pituitary is a gland essential

[1] Any part nearer the head than another part is anterior to the latter; if farther away it is posterior.

to life, and that the anterior portion is more essential than the posterior.

Partial removal of the gland, giving rise to hypo-pituitarism, does not as a rule end fatally, but brings about changes in the animal that in some respects resemble hypo-thyroidism; it becomes fat, it looks dull, and the sexual organs are very imperfectly developed. The animal also shows a greater tolerance for sugars (see the chapter on the pancreas), so that more than the normal quantity of carbohydrate can be taken care of; and the animal converts it into fat and stores it as such. But more characteristic still is the effect on growth. If two pups of the same age and from the same litter are taken, and one has part of its pituitary removed, at the end of twelve months the experimental animal will look less than one-half the size of the control one. This remarkable influence that the pituitary has on the growth of an animal can best be shown with young pups. In the mature animal the other features alone become dominant.

In man, as in the animal, the growth factor is more involved if hypo-pituitarism shows itself before adolescence. The individual remains small and fat; his sex organs show little development; and such secondary sexual features as the formation of hair on the face fails to appear. The temperature is subnormal, the pulse slow. He exhibits a drowsiness and torpidity "like an animal about to undergo hibernation." Sometimes there are psychic derangements; and even epilepsy is not uncommon. Dr. Tucker, in summarizing 200 cases of epilepsy, declares that 63 of this number showed some pituitary disturbance, 28 with an inclination toward the hypo- condition. He further states that feeding these patients with extracts of pituitary gland had a beneficial effect, "not infrequently leading to a cure."

The dwarf and hypo-pituitarism.—Of uncommon interest is the inference that the dwarf is really one who suffers from hypo-pituitarism. While we need much more proof to clinch the matter, some evidence is not wanting in this direction. The few studies that have been made show such pituitaries to be small and atrophied. However, as Professor Cushing says: "It is not unlikely that under the term *ateliosi* and *prageria*, introduced by Gilford to designate continuous youth and 'premature old age', examples of pituitary disease may have been incorporated. Yet Herter describes a type of infantilism which is clearly attributable to chronic intestinal infection; and Osborne and Mendel have shown that feeding young rats with isolated proteins markedly inhibits their growth, though normal weight is maintained.[1] There are still other factors. Hence it is unwise to lay too great stress

[1] See the author's chapter on Amino-Acids in his book on Vitamines.

on anything other than the possibility of an indirect pituitary participation in the dwarfed stature characterizing the many types of infantilism."

Different parts of the pituitary have different properties.—It has recently been shown that in reality the two parts of the pituitary have two very different functions—the anterior part being the one that affects stature, and the posterior portion the one that influences fat formation and the development of the sexual organs. That the anterior fragment of the pituitary is the growth-promoting factor may be inferred from an experiment already cited: if the entire pituitary is removed, death results; if part of it is removed, and a portion of the anterior lobe remains behind, death does not result, but we get a dwarfed condition. To prove that this dwarfed condition of the animal is the result of an insufficient anterior lobe—of insufficient hormone produced by that lobe—and not due to the removal of the posterior portion, the latter alone may be removed, when it will be seen that there is no alteration in stature.

The growth-promoting factor in the pituitary.— Professor Robertson, formerly at the University of California, and now at the University of Adelaide, Australia, has published an extensive series of studies within the past few years on the growth-promoting factor in the anterior lobe of the pituitary. He has isolated a substance from it which he

calls "tethelin" (from the Greek "tethelos," growing), and which he considers the growth-promoting
factor. Its analysis shows it to be related to the
phosphatids, a group of important physiological
compounds present in all cells. These results need
confirmation.[1]

The function of the posterior lobe of the pitui-
tary.—If surgical methods and clinical observation
point to the anterior lobe as the growth-promoting
portion of the pituitary, what, it may be asked, is
the function of the posterior body? We have already seen that surgical removal of the posterior
body is not followed by any extreme changes in the
animal; on the other hand, an extract of the posterior lobe, when injected into the blood, produces
an increase in blood pressure, much like adrenaline
from the adrenal glands (which see). Though
Oliver and Schafer were the first (in 1895) to show
the effects on blood pressure of an injection of an
extract obtained from the *whole* gland, it was left

[1] For the benefit of the student of science the principle that
Robertson employs in isolating his tethelin will be given; undoubtedly much of value is to be found in this pioneer, though
somewhat inconclusive study: The anterior lobe is ground with
anhydrous sodium and calcium sulphates, dried over the water
bath, pulverized and extracted with boiling alcohol, filtered, and
the filtrate evaporated somewhat under reduced pressure, mixed
with one and one-half times its volume of ether, the precipitate
redissolved in alcohol and reprecipitated with ether. The final
product is dried over sulphuric acid at 30 to 35 degrees (cent.).
"From the constancy of its phosphorus and its nitrogen content
the substance would appear to be a chemical unit." Professor
Robertson has not yet ascertained its exact chemical configuration.

to Howell, the physiologist at Johns Hopkins, to prove three years later that the effect was due to the posterior lobe alone.

While there are resemblances between the action of adrenaline and that of the hormone in the posterior lobe of the pituitary, there are also some striking differences. For example, the action of the pituitary hormone is apt to be of a more prolonged nature. Again, a second injection of adrenaline repeats the action of a first injection; a second injection of the pituitary hormone may reverse the action of a first injection. There may actually be a decrease, instead of an increase in blood pressure. Or again, on the arteries of the kidney the two behave diametrically opposite: adrenaline constricts them and pituitary hormone (which we shall refer to as "pituitrin," and to which we shall refer subsequently) dilates them. Pituitrin, indeed, causes quite a remarkable increase in urinary flow, and has established itself as a useful *diuretic* (a substance that increases the secretion of urine).

Properties of pituitrin.—The action of pituitrin on the flow of milk from the mammary gland is no less striking than its property as a diuretic, to which reference has just been made. It does not seem, from the experiments conducted, that the pituitrin actually increases the quantity of milk secreted; rather, it accelerates the discharge once the milk has accumulated. "The pituitrin," says Macleod, "stimulates the muscular fibers of the ducts of the mammary glands, thus squeezing out the milk contained in them."

One other property of this pituitrin must be referred to. We have noticed that partial removal of the pituitary in animals, giving rise to hypo-pituitarism, is followed by an increased capacity for carbohydrate storage (in the shape of fat). We find, on the contrary, that injection of pituitrin decreases sugar tolerance to such an extent as to give rise to sugar in the urine. Evidently it is the lack of a sufficient quantity of this hormone in the animal suffering from hypo-pituitarism that gives rise to an increased sugar tolerance. All of which is in favor of the view that pituitrin, like adrenaline, like the pancreatic hormone (see under pancreas), plays a part in carbohydrate metabolism, though it by no means follows that the three parallel one another in their action.

Pituitrin not a chemically pure product.—We have given the name "pituitrin" to the hormone (or hormones) present in the posterior lobe of the pituitary. It should now be made quite clear that this hormone has not been isolated in the pure state at all; that "pituitrin," of which there are several varieties on the market, is merely a concentrated extract of the posterior portion of the lobe, usually obtained by first getting rid of the protein present in the gland, and then sterilizing the remainder. The fact that sterilization does not destroy the activity of the hormone is further evidence that hormones and vitamines (and enzymes) are probably not very closely related (see also under secretin). The suggestion that the active principle of the posterior lobe is none other than histidine, an amino-acid familiar enough to the organic and physiological chemists [1] (suggested by Professor Abel, of Johns Hopkins, in 1919) has not been confirmed.

Difficulty in interpreting data.—While in a sense our knowledge of the properties of the posterior and anterior lobes of the pituitary is increasing, our difficulty in interpreting data is not diminishing. If the anterior lobe is affected and not the posterior one, we get diminished size and little else. If both are affected, we get a combination of symptoms.

And still another complication arises. As we shall see in a minute, it does not always follow that when the pituitary is affected we get hypo-pituitarism. It may result in hyper-pituitarism—in an excessive formation of pituitary hormone or hormones. Now it is quite conceivable that when one lobe shows such a tendency, another may show the opposite tendency; and yet, since the function of each lobe is different, this does not mean that a

[1] See the chapter on Amino-Acids in the author's book on Vitamines.

neutral or normal condition will result just because the two lobes pull in opposite directions. If we bear these factors in mind—and they probably apply to the thyroid and to other ductless glands— we shall appreciate the tasks of the physiologist and the physician.

Professor Cushing's opinion of the problem.— Professor Cushing in a recent (June, 1921) article on the disorders of the pituitary gland, has this to say: "In the case of pituitary disorders we are not far beyond the stage of the tumor. One may recognize outspoken acromegaly without a roentgenogram of the sella (the place where the pituitary is situated), just as one may recognize exophthalmic goiter without seeing the neck. But in the absence of neighboring pressure signs, to say that a child who is undersized, or fat, or whose dentition or adolescence is delayed, or that an adult who has the texture and color of skin, the adiposity, impotence, subnormal temperature, and so on, known to characterize certain individuals with hypophyseal adenomas (pituitary tumors), is really a subject of pituitary want, is purely a matter of guesswork. If this admission must be made regarding these fairly characteristic symptoms, what is there to say of a pluriglandular (affecting several glands) complex except to acknowledge an abysmal ignorance? . . . Only of late with the development of roentgenology and the more extended use of the ophthalmoscope (a mirror used in examining the interior of the eye) and perimeter (an instrument that measures the field of vision) can a diagnosis of pituitary disease apart from acromegaly be made with any probability. Unlike the thyroid enlargement, a hypophyseal growth can be determined only by indirect methods; for next to the brain itself, the hypophysis lies in possibly the best protected and most inaccessible place in the body—one reason for assuming that it may be a most important member of the endocrine series."

Use of pituitary extract.—Now it is time to answer the question that must arise in the minds of readers: If a person suffering from hypothyroidism, and showing symptoms of myxedema, can be cured by being fed with thyroid extract, is it possible to cure sufferers from hypo-pituitarism by feeding them with pituitary extract? It is in many cases, though one can point to just as many cases where such treatment has been of no avail. Pituitrin extract has been given by mouth and by means of an injection; and a few transplantation experiments have been performed on animals; but the results have been only moderately successful; certainly not nearly as successful as when thyroid extract is administered in the corresponding thyroid disease. Why this should be is not clear, unless we admit that a hypophyseal insufficiency creates at times such a disordered organism, that treatment by means of an extract obtained from any one gland, or even from a number of glands, is no longer sufficient.

Quackery in medicine.—Perhaps in no department of medicine has quackery flourished so much as in that dealing with the ductless glands, and more particularly with the pituitary. The over-enthusiastic scientist has joined hands with the pseudo-scientist, and both have provided ample material to the charlatan to advertise his goods, and to ingratiate himself with an all-too-credulous public.

I cannot resist quoting Professor Cushing's very pertinent remarks: "Children are either too short or too tall, too fat or too lean. Their adolescence is too early or too late; they have too little or too much hair. They are intellectually backward or stupid, even defective or epileptic . . . all this needs attention and can be corrected by some whole-gland extract, usually with a pinch of thyroid thrown in.

"Pituitary extract is advocated in parturition, shock, baldness, impotence, epilepsy, and a multitude of other conditions which have hitherto baffled us; and if it does not suffice by itself you are earnestly recommended, according to the directions in the enclosed folder, to try this or that combination of hormones which contains the active principle of several glands . . .

"I know of nothing comparable to the present

furore regarding the administration of glandular extract unless it be the plant pharmacology of the middle ages. . . . A patient is bilious—therefore he has some disease of the liver. The leaves of a certain plant resemble in their color and appearance the surface of the liver—therefore a concoction of these leaves is good for biliousness, and the plant comes to be called hepatica. But then, lest it may not really do this, we will add several other things to the concoction as well. This is about the basis on which the glandular extracts are administered to-day. And it will be noted that most of them contain a certain amount of thyroid extract, which possibly is the only one of these substances having any definite action when given by mouth. . . .

"Surely nothing will discredit the subject so effectively as pseudo-scientific reports which find their way from the medical press into advertising leaflets, where, cleverly intermixed with abstracts from researches of actual value, the administration of pluriglandular compounds is promiscuously advocated for a multitude of symptoms, real and fictitious. The Lewis Carroll of to-day would have Alice nibble from a pituitary mushroom in her left hand and a lutein (a pigment obtained from a portion of the ovary) one in her right hand and presto! she is any height desired!"

Pituitary extract and the development of chick-

ens.—One rather remarkable experiment with chickens, performed by Dr. L. N. Clark, must be cited: "In the first experiment, 35 white Leghorn hens, as well as two cockerels of the same breed with which they were mated, each received daily during eight days the equivalent of 20 milligrams (0.0006 ounces) of fresh pituitary substance in addition to their usual food. By the fifth day the egg production of the batch was raised from an average of 18 per diem to 33; the beneficial effect, although diminishing, was maintained for several days after the pituitary had been taken off. And not only was the output of 'eggs largely 'increased as compared with the controls, but the fertility of the eggs and the hatching out of the chicks was extraordinarily enhanced. In order to test the matter further, a second experiment was performed with as many as 655 one-year-old white Leghorn hens (kept without males), the same dose as before being given to each hen during four days. The average daily number of eggs laid by the batch during the four days preceding the pituitary feeding was 233; during the four days succeeding the administration, 352. These experiments were made at a time of the year when the egg production of the hens was tending to diminish rather than to increase." (Quoted by Professor Schafer.)

Hyper-pituitarism.—Marie, a French physician, active some forty years ago, was the first to give the name "acromegaly" (enlargement of the ex-tremities) to that disease which produces an en-largement of bones, soft part of hands and feet, and face, and in which the individual assumes giant-like proportions. He rightly diagnosed such cases as being due to some pituitary disorder, but for some time he was of the opinion that acromegaly was due to lack of secretion in the gland. When, however, experiments on animals showed that the partial removal of the gland gave rise not to giants but to dwarfs, the view was adopted that acrome-galy was a disease resulting from a hyper-, rather than a hypo-secretion; that is to say, then, to an excessive secretion of pituitary hormone. It is un-fortunate in the extreme that so far all attempts to simulate acromegaly by feeding animals with large quantities of pituitary extract have failed; so that our evidence at best is somewhat unsatisfactory. The reader will remember that in this respect thy-roid therapy stands on firmer footing; for hyper-thyroidism can be fairly closely imitated in the ani-mal by feeding it with sufficient quantities of thy-roid extract.

Symptoms in acromegaly.—The symptoms in acromegaly do not make their appearance sud-denly; the changes in the individual are quite gradual, as with many of the diseases involving the ductless glands. At first the person may notice nothing more than eye strain and dimmed vision.

This may be due to a tumor in the gland pressing on the optic nerve. But then comes the gradual enlargement of the head and extremities that can leave no further room for doubt. The face becomes big and coarse-looking, the head rather flattened. An X-ray examination of the fingers shows the bones to be enlarged. The entire skeleton increases in size, sometimes to huge proportions. Then also we have an overgrowth of hair, underdeveloped sexual glands, and a probable onset of diabetes,—which may, however, be of a temporary nature.

Cure in acromegaly.—We have no easy cure for this disease. Surgical interference may seem an obvious remedy. This is a dangerous procedure, however, and is undertaken only in extreme cases. Fortunately, true cases of acromegaly are rare, and experience has shown that nature and time often act as effective curative agents.

As a rule the case presented to the physician and surgeon is by no means a clear-cut one. It may be one where hyper-pituitarism predominates, giving us an example of gigantism if the patient is affected early in life, or acromegaly—not so pronounced as gigantism—if later in life; or it may belong to the hypo-pituitary class, with reverse symptoms, such as under-size, adiposity and sexual infantilism, if early in life; or the latter two, if later. But a far more common case, and one that complicates the situation, is the mixed or transition type, exhibiting some features of the hyper- *and* the hypo- conditions (*dys*-pituitarism).

Of 255 cases that have come under Professor Cushing's care, 200 showed evidence of a tumor—which of itself may give rise to a hyper- or hypo-condition, or to a mixture of both. Of these 200, 180 were operated on, and among the latter there were few acromegalics. The majority exhibited various forms of dyspituitarism.

Professor Cushing writes: "In view of the fact that hyper-pituitarism, so far as glandular secretion is concerned, is a condition which tends to right itself, it must remain for the time being a matter of uncertainty as to whether or not in the absence of a degree of hyperplasia (excessive multiplication of tissue elements) sufficient to cause neighborhood symptoms, operative measures can hold out any promise of permanently controlling the disorder. When, however, neighborhood symptoms have arisen owing to the extreme enlargement of the gland, due to the formation of an adinomatous struma (a tumor in an enlarged gland), whether or not there have been antecedent symptoms of acromegaly, the surgical aspects of the matter stand on firmer ground."

Do giants suffer from hyper-pituitarism?—One is attracted by the hypothesis that giants are examples of men suffering from hyper-pituitarism.

Arthur Keith, the sponsor of attractive theories on behalf of the endocrine glands, writes, "In giants this gland [the pituitary] is always greatly and abnormally overgrown. We have reasons for supposing that it has thrown a drug or drugs into the blood which set all the bone-builders into a state of frenzied activity, producing the dire disease of growth called gigantism." These claims cannot be maintained without many qualifications.

A case of acromegaly.—Of a number of instructive cases described by Cushing in his book on the pituitary—a book that has become a classic in medical literature—I shall cite one as illustrating the acromegalic who in time actually suffers from hypo-pituitarism—in other words, a complex type.

The case was that of a farmer, age 35, of Dutch extraction and of excellent family history. A hypophyseal tumor caused pronounced neighborhood symptoms with almost total blindness. There had been a former glandular activity, shown by a tendency to excessive overgrowth, traceable to the adolescent period; and by subsequent addition of acromegalic changes of unusual degree. When admitted to the hospital the patient, on the contrary, indicated a condition of glandular insufficiency, as shown by adiposity and high sugar tolerance.

As to his history, nothing unusual was noticed in the farmer until he was 13 years old, when he began to grow with extreme rapidity, and at 19 measured about six feet four inches, having developed into a powerful man of unusual strength, weighing 222 pounds. He was intelligent and a good student. So far all had gone well. The man was an excellent type of a physically well-developed youth.

At 23 he had a severe illness; he was said to be "threatened with consumption." At 25 there seem to have been no traces of acromegaly. He and his father were positive that a second growth began when he was 27.

About 1903—at 28—he began to have violent headaches (due to a tumor of the pituitary, and to the consequent pressure on adjacent parts); also pains in the extremities. These attacks would be followed by the discharge from the nose of quantities of "slimy mucus," occasionally tinged with blood, and relief would ensue for some days or weeks. He was told at the time that he had acromegaly. Two years later difficulty in sighting his rifle first called attention to a failure of vision.

In 1907 his parents realized that his "features were changing," and that he was "getting large all over," and was losing his strength (this was during his "second period of growth").

When admitted into Dr. Cushing's clinic—in 1910—he had become very weak and drowsy, and tired easily. There had been a complete loss of *libido et potentio sexualis.* Reading vision was lost the year before. The left eye was blind and the right nearly so. His appetite was large and he always had a box of sweets at hand. He had become very constipated.

The patient's height was six feet six inches. He weighed 269 pounds. "Neither these measurements nor the photographs give more than a scant indication of his extraordinary size and the disproportion of such parts as the head, hands and feet. He is a veritable Gargantua."

The X-ray indicated tumor enlargement of the pituitary. The eyes were large, protruding, widely separated, and showed a divergent squint (neighboring symptom).

The hands, always large, were now huge. The feet were colossal: formerly he could wear a number eleven shoe, but now he had to have them specially made.

He had practically no beard, and except for a scant pubic growth of feminine distribution, the skin of the trunk and extremities was practically hairless. On the scalp the hair was abundant and coarse.

The patient was operated on nine days after admission, and the glandular tumor thereby removed. The convalescence was uneventful.

On December 23, six days after the operation, there was an improvement in vision.

Since the later symptoms were those of hypo-

pituitarism as well, glandular feeding was instituted.

On January 28 the patient felt so much stronger
and less nervous, that he was discharged.

A letter received from the patient February 27
reported a further improvement in vision, a normal
temperature, and no constipation. On May 4 he
reported an increase in weight to 281 pounds (just
before the operation this weight was 269). On
June 27 the patient was seen in Los Angeles. The
vision in the right eye had further improved. He
recognized colors. He was less nervous and less
drowsy than before.

A bit of fancy?—At the International Congress
of Eugenics, held in New York in September, 1921,
a session was given over to the influence of the
endocrines on the organism. The papers presented
were of the type that belong to the borderland sepa-
rating fact from fancy. At any rate, one of New
York's very respectable papers had headlines such
as these: "Says glands cause gloom and crime.
A criminal is the victim of chemical reactions.
Scientist explains why a white man is white and
why man is superior to woman." One of the papers
dealt with the development of man from the monkey, and Professor Polk, Director of the Department of Anatomy in the University of Amsterdam,
discussed the first change in man's ape ancestor,—
the suppression of his hairy covering. He argued

that this suppression was connected with pituitary activity! It would seem that Darwin in developing his theory of Natural Selection, quite overlooked the subject of endocrinology!

CHAPTER V

THE ADRENAL GLANDS

These glands consist of two small bodies situated near the kidney (hence "adrenal"), each weighing about one-seventh of an ounce. Sometimes they are called "suprarenal bodies," to indicate that they are found *above* the kidney. Sometimes they are referred to as part of the "chromaffin system," to indicate that their cells are colored brown with chromic acid.

There are really two very distinct parts to the adrenal gland: the "medullary matter," a marrow-like body, in which is found the hormone adrenaline (also known as epinephrine and suprarenine), and the outer organ, the "cortex," which also secretes a vitally necessary substance that so far, however, has baffled isolation.

History.—Eustachius, the great anatomist of the sixteenth century, may be regarded as the discoverer of the adrenal glands. As early as 1789 Cassan made the observation that the adrenals of the Negro are larger than those of the European, from which he drew the somewhat far-fetched conclusion that the gland and the pigment of the skin are related. In a modified form, this view of Cassan's was recently brought forward at the International Congress of Eugenics. Meckel, another observer, noticed a similar enlargement of the gland in the Negro, but he decided that the relationship did not rest with the pigment of the skin, but rather with the genital organs.

Little was known with regard to the function of the adrenals until 1849, when Thomas Addison, physician at Guy's Hospital, London, made the observation that a disease characterized by bronzing or pigmentation of the skin, was invariably accompanied by the decay of the adrenal glands. This observation of the English physician led to little new work, however; and it was only as late as 1894 that a further impetus was given to the entire subject by Professor Schafer's discovery that the injection into the body of an adrenal extract increased the blood pressure.

Removal of the gland.—Complete removal of the gland in animals is followed by death within a few days, though the first day or two after the operation may fail to show any abnormality or disease. Partial removal is not as a rule fatal, though various symptoms may make their appearance. The administration of adrenal extract has little or no effect. Even in the case of Addison's disease, which we have reason to believe is due to a decreased secretion of the gland, feeding with an extract of the gland is followed by no beneficial results—at least none that last.

Brown-Séquard, whose name we shall encounter when we come to discuss the function of the sex glands, did pioneer work on the adrenals as far back as 1856. He removed adrenal glands from 44 rabbits, nine guinea-pigs, two rats and several dogs and cats. Not one of these animals survived the operation beyond the thirty-seventh hour. The critics, ever after poor Brown-Séquard, severely criticized his work; they claimed that death in all cases was due to the nature of the operation, and not to the removal of the gland. Brown-Séquard repeated and extended his observations. His conclusions are shared by the majority of physiologists to-day. Even his claim that partial extirpation does not as a rule result in death has been amply verified. In his over-enthusiasm—and Brown-Séquard was a most enthusiastic gentleman—he declared that the adrenals were even more important to well-being than the kidneys themselves.

Not the least remarkable thing to be noticed in these extirpation experiments is the way the animal will appear quite normal for some time after the operation, and then quite suddenly begin to exhibit effects. Let us illustrate this by giving the protocol in the case of a female Macacus monkey, whose adrenals were removed by Dr. Kahn (quoted by Mathews). The animal weighed a little over four pounds and fed on fruit. "Right suprarenal removed under ether Nov. 9, 1911. The wound heals quickly. On 4 December, 25 days after the operation, when the weight was 62 ounces, took out the left suprarenal. On the fifth the animal is very well and eats heartily. On the sixth she eats with normal appetite. Is active. There is a little edema (swelling) at the edges of the wound. On the seventh, normal. On the eighth, normal, but appetite a little less. On the ninth at 9 a.m. is fairly weak, lies stretched out on the bottom of the cage; no appetite; wound in best state. At 9:12 a.m. great increase in prostration. Apathetic. The eyes are open and look about. At 1:45 p.m., being nearly moribund, was killed with chloroform. The liver was examined for glycogen; only a trace was found. The animal lived five days. *For four days it could not be told from a normal animal.* The sudden onset of the symptoms of extreme depression has the appearance of an intoxication."

Why the "normalcy" for some days after the operation, and then the sudden change for the worse, is not at all clear. Perhaps the body has a hormone reserve that holds out for four days and then gives out.

Addison's disease.—The equivalent—or what is supposed to be the equivalent—of the removal of the adrenals may be seen in man in Addison's disease. Dr. Addison regarded the disease as "asthe-nia (loss of strength), irritability of the stomach and change of color in the skin." In a recent (Feb., 1921) issue of the *New York Medical Journal*, Dr. Eidelsberg records a case; let us see how the symptoms agree with Addison's. The man, aged 30 years, was admitted to the medical ward of the Post-Graduate Hospital, complaining of abdominal pain, vomiting, and weakness. The illness began suddenly ten days previously. There was loss of appetite, weakness, vomiting immediately after each meal, constant pain in the abdomen, marked constipation, dizziness, and fainting sensation on slightest exertion. These symptoms continued throughout the entire ten days, increasing in severity. The patient appeared "tanned," as though he had exposed himself to the sun's rays. There were about 15 small, pale scars scattered over his body. The weakness rapidly increased until the patient could not turn in bed. The vomiting grew worse, and no food was retained.

On the evening of the sixth day after admission to the hospital, the patient suddenly became unconscious, his pulse became imperceptible, and in five minutes he ceased to breathe. An autopsy examination revealed that both suprarenals were markedly enlarged and soft, and on cutting, appeared, except for a thin cortical portion, to consist of a soft, cheesy material. Several areas typical of a tuberculosis lesion were found.

The "pigmentation" is usually a characteristic feature of the disease. Sometimes it is absent, and sometimes it is present in diseases other than Addison's. Where it is present the color may vary anywhere from the dark hue of the Negro to a faint sunburn brown. Characteristic also of Addison's disease are the signs of muscular weakness (with no corresponding emaciation). "The patient is very easily tired, and is never able to get properly rested." One may add that in most cases the blood pressure is very low, the heart feeble in action, the temperature is usually subnormal, and the patient presents an anemic appearance.

The marked pigmentation of the skin in Addison's disease is supposed to be due to a disturbed relationship between adrenaline and melanin, the characteristic pigment present in the skin. When the adrenals are diseased and the amount of adrenaline reduced as a consequence, the factor that controls melanin formation is removed, and an excess of the skin pigment is deposited.

The reader will notice analogies between this theory and the one advocated by Cassan more than a hundred years ago.

Effect of adrenal extract.—We have seen that cases of hypothyroidism, such as myxedema, can be cured by administering thyroid extract; can we cure an analogous disease of the adrenals by the administration of adrenal extract? Can we relieve patients suffering from Addison's disease by any such method? We have already indicated that the answer must be in the negative. We can supply no good reason for the failure, unless we assume that the adrenal hormones, unlike the thyroid ones, are quickly destroyed in the system; and even then a veil of obscurity still overshadows the situation. We must remember in this connection, as pointing to how far from a complete solution we really are, that the most characteristic feature of Addison's disease, the pigmentation of the skin, has never been experimentally produced.

"Grafting."—From what has been, and what will be said on the subject of grafting (see more particularly Chapter VI), one might suppose that the use of such a method would be an improvement over the use of extracts. If the graft takes and a circulatory system is set up, the gland ought to behave like any organ of the body that is "alive" and that functions properly. Theoretically it ought to do so; in practice it seldom does. Surgeons in the late war have shown much ingenuity in grafting pieces of skin; occasionally thyroid grafts have been carried out successfully; but when we come to adrenal grafting we can record little else but failure. The gland atrophies and the med-

ullary portion disappears altogether. Some experimenters have recorded temporary successes, but nothing that as yet warrants much optimism. Jaboulay, a French surgeon, transplanted the adrenals of a dog into a patient suffering from Addison's disease. The surgeon does not tell us what happened to the transplanted gland, but he does record that the patient died within 24 hours! Dr. Voronoff might reply that one of two reasons would explain this: either the surgeon was a bungler and did not perform the operation skillfully enough; or, what is more probable, the adrenal of a dog cannot replace that of man. Biology and chemistry certainly have taught us that "specificity" is a distinguishing feature of many dynamic reactions. In any case, before we condemn adrenal grafting, many more such experiments will have to be performed.

Hyper-adrenalism.—Neither is our information less obscure with regard to conditions where there is an excessive production of the adrenal hormone —here again in striking contrast to our knowledge of hyper-thyroidism. "The whole subject," writes Vincent, an eminent English authority, "is very obscure and requires further and continuous investigation."

The cortex.—To what extent our ignorance of the entire subject of the adrenal glands is due to our ignorance of the part played by the cortex of the gland, it would be hard to say. You will remember that at the beginning of the chapter we stated that the gland consists of two very distinct portions, the cortex and the medulla. The latter, as we shall see, contains adrenaline, the best known of all the hormones, and indeed a very important substance. But evidence is accumulating to show that the medulla, which contains this hormone, is no more important, if indeed as important, to life than is the cortex, which does not contain adrenaline. Biedl, a celebrated Austrian investigator, claims to have succeeded in removing the cortex from mammals, leaving behind the medulla intact; the animals did not survive. This leads him to the view that the cortex, and not the medulla, is the portion of the organ essential to life. Schafer, the Edinburgh physiologist, has criticized Biedl's conclusions because he considers it impossible to separate completely the medulla from the cortex. "I think," he writes, "the experience of most people will lead them to believe such a separation impossible." Yet the view does persist that the cortex, if anything, is even more important to the body than the medulla.

What hormone, if any, the cortex contains, is not clear. No substance corresponding to the adrenaline of the medulla has been isolated from it. A theory has, however, been advanced that the real seat for the manufacture of adrenaline is the cortex; that the hormone is there made from tyrosine, one of the best known of our amino-acids, which in turn are decomposition products of proteins;[1] and that the adrenaline, once formed, is passed to the medulla, where it is stored. This is an attractive hypothesis, but, like many attractive hypotheses, lacks experimental proof.

Another view of the function of the cortex is that it destroys poisons—either those produced in the course of body metabolism, due perhaps to muscular activity (theory of auto-intoxication), or those entering the body from the outside. Still another theory is the one that stresses the close relationship existing between the cortex on the one hand, and the generative glands on the other. It has been said that the enlargement (hypertrophy) of the adrenal gland goes hand in hand with precocious development of the reproductive organs.

The medulla.—When we come to discuss the function of the medulla we are on much firmer ground, for here we shall see we have a tissue that contains a substance which itself induces some remarkable changes. We must now discuss this substance—adrenaline.

Adrenaline.[2]—The most significant advance made in our study of the adrenals is the isolation, in a

[1] See the chapter on Amino-Acids in the author's book on Vitamines.
[2] The relation of adrenaline to the nervous system is reserved for Chapter XII.

chemically pure state, of one of its hormones, adrenaline. This substance has not only been isolated from the gland, but it has also been synthesized in the laboratory. The organic chemist gives to it the name orthodioxy phenyl-ethanol-methylamine, and he knows that it is therefore related to tyrosine, an important amino-acid obtained when proteins are decomposed.[1] With the help of a little fancy and a little fact, the physiological chemist explains the indispensability of some of these amino-acids by declaring that they constitute the raw materials for the manufacture of hormones.

If we except the isolation of the hormone from the thyroid, the isolation of adrenaline remains the only case on record of the extraction of a hormone in a pure condition from its gland. "It is one of the greatest triumphs of physiological chemistry," writes Vincent, "that within seven years of the discovery of the powerful effects of extracts of the adrenal medulla by Oliver and Schafer (Schafer's discovery was made in 1894), the active principle was obtained in crystalline form, and that five years later its composition has been so completely ascertained that it has been synthesized, and the pure active synthetic products can be obtained from the manufacturing chemists."

The men primarily responsible for the isolation

[1] See the chapter on Amino-Acids in the author's book on Vitamines.

of adrenaline from the adrenals are von Fürth, an Austrian; Abel, professor at Johns Hopkins; and Takamine, a Japanese domiciled in the United States. Friedmann, a German, succeeded in producing adrenaline synthetically in the chemist's laboratory. While it is not my intention in a volume of this kind to enter into any details regarding the chemical steps involved in the isolation of hormones, the briefest outline of Abel and Takamine's methods for isolating adrenaline from its gland will be given. Even though such an outline proves too "technical," it can hardly be omitted from a book purporting to deal with hormone action.

Preparation of adrenaline.—Very concentrated adrenal extracts are largely freed from inactive substances by treatment with alcohol, lead acetate, etc.; then the active substance is precipitated in microscopic crystals by the addition of concentrated ammonia. The precipitate is then purified by repeatedly dissolving in acid and reprecipitating with ammonia. The resulting prismatic needles or rhombic plates are those of the purified or isolated active principle—adrenaline.[1]

[1] For the benefit of students of chemistry who may read this volume, a word may be added as to the synthetic production of adrenaline. This may be obtained by the action of methylamine upon chloroacetopyrocatechin:

$$C_6H_3(OH)_2COCH_2Cl + NH_2CH_3 \rightarrow C_6H_3(OH)_2 . COCH_2 . NHCH_3 . HCl$$

The methylamino aceto-pyrocatechin so obtained yields adrenaline on reduction.

Properties of adrenaline.—Before entering upon a discussion as to the various uses that adrenaline is put to, it may be said that, in a general way, adrenaline affects the body tissues in much the way that the sympathetic nervous system does. By way of further explanation it should be remarked that the nervous apparatus consists of two sets of nerves connected, to be sure, and yet standing out apart from each other: the cerebro-spinal system and the sympathetic system. The former is made up of the brain and the spinal cord with its corresponding (cranial and spinal) nerves; the latter consists of a chain of nerve cells extending on each side of the spinal column, connected with each other and with the spinal nerves. From the sympathetic system, nerves generate that follow to a large extent the distribution of the blood vessels, and that form large networks around the heart, stomach, etc. This system controls the internal organs, such as the heart, the blood vessels, kidneys, etc., and the sweat glands and vessels of the skin; it controls involuntary muscular movement, and is related to

The synthetic adrenaline is optically inactive,—that is, it is the *d-l-* adrenaline; whereas that obtained from the adrenals is the optically active *l*-adrenaline. However, not only has the inactive or "racemic" mixture been separated into its active components, but the interesting fact has been brought out by Abderhalden that the *l*-adrenaline is about 15 times as strong in its action on blood pressure as the *d*-adrenaline.

For a fascinating account of optically active substances, the general reader can do no better than read Pasteur's investigation of tartaric acid, as related by his son-in-law, Valery-Radot, in his book dealing with the life of that immortal French scientist.

such processes as contraction and dilation, secretion, and various nutritional processes.

In their action the sympathetic nerves are often antagonistic to the cranial ones. The stimulation of the latter stops the heart beat; the stimulation of the sympathetic fibers quickens the heart beat. So, indeed, does adrenaline: its action is to quicken the heart beat.

It would be difficult to name a substance used in medicine that has proved of greater value than adrenaline. It causes a contraction of the arteries and is an excellent hemostatic,—that is, it checks the flow of blood. In shock and collapse, often following surgical operations, the procedure adopted by Crile, the well-known Cleveland surgeon, of administering adrenaline solutions, has been generally adopted. Though we know little as to the cause and nature of shock,[1] we do know that in such a condition the arteries contain less blood, and the veins more blood than usual. Since adrenaline constricts the arteries, its use under these conditions provokes attempts on the part of the body to restore equilibrium.

The suggestion has been made that trench life in war leads to depression of adrenal activity, and that this is connected with the clinical picture of shock in war. We shall return to this later.

It might be expected that cases of hemorrhage,

[1] For further information see Chapter XII.

with the copious loss of blood that accompanies them, would be benefited by adrenaline treatment. This is in fact true of many such cases. Injection of the hormone constricts the blood vessels, and, as Professor Cannon, of Harvard, has shown, actually hastens the formation of a blood clot, which in turn acts as a seal to any further escape of blood.

The extensive use of adrenaline in conjunction with a local anesthetic dates from 1903, when Braun, a German surgeon, found that a subcutaneous injection of the hormone produces a bloodless or bleached area—of great importance to the surgeon in giving him a "clear field of operation"— even better than the method of bandaging or freezing. Then, as a further incentive to its use, it was shown how, applied in conjunction with cocaine or novocaine, the effects of the anesthetic are increased and last longer; neither are there such unpleasant after-effects. Since then adrenaline in conjunction with anesthetics has been used more and more.

One may cite its invariable use in ophthalmological surgery. In eye operations and examinations the application of adrenaline has opened up many new possibilities.

Since we are on the subject of the eye, it may be of interest to record that in a state of mental disorder which goes under the name of "dementia precox," it has, at times, been found possible to differentiate this disease from others showing certain similarities by injecting the hormone into the membrane that lines the eyelid and covers the eyeball (the "conjunctiva"); after a few minutes there follows a marked dilatation of the pupil ("midriasis"), provided the patient has true "dementia precox."

The amount of adrenaline used in these operations and tests is ridiculously small; perhaps as little as five to ten drops of a solution containing one part of hormone to 1,000 parts of water. Small as such a quantity is, it is yet strikingly effective. What is true of adrenaline seems to be true of all hormones: that their effect is altogether out of proportion to the small amounts of material used. It is this disproportion between amount of material used and the profound changes brought about by such a quantity, that has caused investigators to compare hormones with vitamines and enzymes. What connection there is between them remains a mystery.

Adrenaline in blood.—We know that even under normal conditions the blood always contains a small quantity of adrenaline—so small, indeed, that it has been estimated at one part in 500 million of blood. For reasons that will soon be apparent, much attention has been given to methods for the quantitative determination of adrenaline. It must be remembered that the quantities involved are almost infinitesimal in amount, and to devise not only qualitative, but quantitative methods for detecting such amounts, appears to be a problem beset with endless difficulties. Yet what seems to approach the impossible has been accomplished. In this connection it is of interest to note that physiological methods, like spectroscopic methods in many cases, and the electroscopic method in the case of radium, have proved themselves more delicate than chemical ones. For example, Meltzer has devised a physiological procedure which depends upon the action of adrenaline in dilating the pupil of the enucleated eye of a frog; and claims that a strength of hormone one part in 20 million can be detected. Using a loop of intestine and recording how its movements are inhibited by adrenaline, is said to be so delicate as to detect one part of hormone in 400 million.

The best chemical method so far devised is one based on colorimetric comparisons, due to Professor Folin, of Harvard. It depends on the blue color obtained when adrenaline is added to phosphotungstic acid. This blue color can be noticed in dilutions of adrenaline of one part to three million.

Adrenaline and sugar metabolism.—The relation of adrenaline to sugar metabolism will be referred to in Chapter VII. Here it should be said that the injection of this substance into the body of an animal increases the quantity of sugar normally found in blood, and may give rise to a temporary glycosuria (sugar in the urine), which is a form of what is commonly called "diabetes." It will be shown that the pancreatic hormone acts antagonistically to the adrenal hormone, thereby helping to regulate body equilibrium.

Professor Cannon has shown that an increase of adrenaline in the blood occurs without any addition of hormone from outside sources whenever an animal gets excited—for example, when a cat gets ready to defend itself against a dog's attack; and also that such an increase is followed by a hyperglycemia (an increase of sugar in the blood). This suggestion that the adrenal gland is involved in emotional outbursts is certainly of extraordinary interest. The fact remains too that when a cat becomes frightened its "pupils dilate, the stomach and intestines are inhibited, the heart beats rapidly, the hairs of the back and tail stand erect—all signs of nervous discharges along the sympathetic paths"; and we have already seen how similar the action of adrenaline and the stimulation of the sympathetic fibers are.[1]

Summary.— We may briefly summarize our knowledge of the adrenals by stating that they are unquestionably essential to life, though conflicting theories are current as to their precise function. Of the two parts into which the adrenals may be

[1] See further, Chapter XII.

divided, the cortex and the medulla, the former, if anything, seems to be more essential to life than the latter, though so far no one has succeeded in showing that the cortex contains any specific hormone. On the other hand, the medulla contains the best known of all the hormones, adrenaline. This hormone, adrenaline, has not only been isolated in the pure state, but it has actually been synthesized in the laboratory. Adrenaline has already taken its place among the most valuable drugs in medicine.

CHAPTER VI

THE ORGANS OF REPRODUCTION

We have convincing evidence that the organs of reproduction, or the "sexual" glands, produce both an external [1] and an internal secretion; in this respect showing a similarity to the pancreas and the small intestine. The external secretion contributes to the reproduction of the species; the internal secretion—like the secretions from the other ductless glands that we have studied, contributes to the molding of the species. In this chapter we are, of course, primarily interested in the internal secretion of the sexual glands, though the two secretions cannot always be sharply separated.

Proof of an internal secretion.—The removal of the testes or ovaries, an operation commonly known as "castration," if carried out before puberty, prevents the development of "secondary sex characteristics." In the female the periodic act of menstruation from a very early age until perhaps her fiftieth year, is indicative of a functional ovary. The

---

[1] Some physiologists are opposed to regarding the external secretion of the reproductive system as a true secretion, on the ground that the active constituents are not enzymes.

"coming on of heat" or "rut" in animals seems to be related to menstruation in human females. Now it is a remarkable fact that this "rut" is no longer noticed upon the extirpation of the ovary, but is again brought on by the transplantation to another part of the body of the ovary belonging to a similar animal. From what has already been said on the subject of transplantation, it must be perfectly evident that an experiment of this kind admits of but one conclusion: that the nervous mechanism plays but a secondary part, if it plays a part at all; but that on the other hand, the ovary must discharge into the blood some substance or substances which give rise to the phenomenon in question. This, of course, means that the ovary, in addition to giving rise to an external secretion in the form of ova, that contributes to the reproduction of the species, must be the seat of an internal secretion.

We now know that the two types of secretion are produced by two different types of cells. The internal secretion is developed by the so-called "interstitial cells," the name of which has become familiar to the lay reader ever since topics on rejuvenation have become popular with newspaper editors. These cells lie altogether outside of the tubes that are responsible for the flow of the external secretion. It is, in fact, possible to tie the tube that, in the male, connects the testes with the ejaculatory tube (the "vas deferens"), thereby causing the sexual elements to disappear; yet the interstitial cells not only are left, but actually increase in number. And what is far more remarkable, the sexual instincts remain as strong as ever. This is therefore further evidence in favor of the view that puberty is dependent not upon the presence of the reproductive elements, but rather upon the production of an internal secretion due to interstitial cells.

Brown-Séquard.—In 1889 Brown-Séquard, one-time professor at Harvard, and later professor at the Collège de France in Paris, presented a paper before the *Académie de Médecine* that created a sensation. He described experiments in which he had himself injected with extracts obtained from the sexual organs of a ram; the result was that he had become quite "rejuvenated." Despite his seventy years he felt, he said, like a youth, with all a youth's vigor. The newspapers, quite justifiably, gave much prominence to this get-young-quick method, and for a short time Brown-Séquard was an international hero. Then came cries from opposition forces. The experiments were repeated and nothing like what was alleged by the French-American could be discovered. Brown-Séquard answered his critics by charging that his technique had not been followed with sufficient care. But his replies did not satisfy the critics; they attacked him mercilessly, and continued the attack until the one-time famous physiologist became a thoroughly discredited man. We shall presently have occasion to discuss the objections that may be advanced to the use of testicular extracts; but it is only bare justice to the memory of an illustrious man of science to point out at once and emphatically that Brown-Séquard seized upon the germ of a great discovery, and one which, even to-day, is but dimly perceived. Moreover, he was the first one to perceive clearly the intimate relation that exists between the various organs, due to their internal secretions.

He writes: "Each tissue, and, more generally, each cell of the organism, secretes for its own use special products which are poured into the blood and which influence, through the intermediary agency of this liquid, *and not through the mechanism of the nervous system,* all the other cells, thus rendering all of them mutually interdependent." To say that each tissue and each cell plays such a part may, or may not be the entire truth; but we have here a wonderfully clear presentation of the functions of hormones.

We shall see presently how, in the hands of Steinach, Voronoff, and others, the Brown-Séquard view was adopted and extended. In the meantime we must take a step backward and examine in more detail the effects of castration. You will remember that this type of approach to the subject is the one that so often yields valuable clues as to the function of any particular part of the organism. *Castration.*—Here the history of mankind supplies us with an abundance of material—at least, in so far as the male part of the population is concerned; for our information concerning the female portion is slender. Castration was extensively practised in antiquity, and still is in oriental countries where there are watchers of the harem. Professor Falta informs us that it was carried out in Italy "for musical purposes," and that the practice ceased only recently. In Russia a religious sect known as the "skopzen" include castration as a necessary part of their religious ritual. Perhaps the most pathetic account that has been handed down to us is that of the famous Abélard, who was castrated by order of Héloise's uncle, Fulbert.

The results of castration are much dependent upon the age at which it is performed. If this is done before puberty, the sexual instinct disappears never to return. The eunuch often grows to be very tall, those reaching six feet eight inches being common. The voice, due to the non-development of the larynx, retains the sound of a childish soprano. The face is livid, the skin poor in pigment and the flesh flabby. Very often the breasts are abnormally developed, and almost always such individuals are beardless. These last two symptoms have often been taken as an indication of the development of female characteristics, and as showing that the sex characters owe their presence to the genital organs. The very name suggests that the sexual organs are involved in the development of sex characters; but why when such organs of a male are cut out of his system he should then tend to revert to the female type is not clear. The view is gaining ground that castration produces a "neutral" rather than a female variety.

What effect castration has upon mental development is not beyond dispute. From the accounts of some eunuchs that have come down to us, who attained distinction in their day, it would seem as if the mind of the person is not affected. Unfortunately our information as to their age at the time of castration, and as to the completeness of the operation, is meager; and both these factors need to be known before we pass judgment. Experiments with animals have shown that castrated animals lack the courage, the animation and the passions of normal male animals; and many observations on eunuchs lead to the belief that there is a distinct tendency towards the suppression of the finer emotions, these in turn being supplanted by a general air of indifference and mental inertia.

The absence of certain sex characteristics—which does not always imply the formation of sex characters belonging to the opposite sex—after castration is also noticeable in animals. Thus stags when castrated young do not develop antlers; and even if the operation is performed after the antlers have appeared, the animals no longer shed them annually. If horns are on they fall off. This is true of sheep. The cock loses its plumage, its wattles and its characteristic male type of voice.

The records of castration in the female are meager. The operation is rarely performed before puberty, and after this stage has been reached, the changes are not so marked. This applies also to domestic animals. The few experiments that have been performed make it seem likely that one of the functions of the female genital organs is to suppress a tendency to revert to the male type; in their absence the duck and pheasant, for example, assume male plumage.

In both the male and female, whether man or animal, castration is followed by an accumulation of fat in the body, making it appear that the general metabolism of the system has been disturbed.

The results of castration can be observed in those persons whose sexual organs have for one reason or another not developed normally. Professor Tandler refers to such "eunuchoids" as "tall, or if complications are absent, at least not stunted in growth; they show the typical fat distribution of eunuchs. The skeletal dimensions are characterized by an especial length of extremities. There is a more or less pronounced disturbance of the development of the genitalia, with faulty development of secondary sex characteristics. It is probable that in such cases we have to do with a developmental disturbance beginning primarily in the sexual glands, and indeed the interstitial glands, as functional disturbances of the generative glands alone do not lead to eunuchoidism."

The last statement receives support from experiments already described, wherein the reproductive elements were suppressed, yet the gland consisting of the interstitial cells thrived, and the several sex characteristics were retained; also where a tumor arises that interferes with the supply of the reproductive elements, yet does not attack the interstitial cells, we may get a similar result. On the other hand, where the interstitial cells themselves are affected, the clinical picture obtained represents that of castration; which in itself suggests that the primary results of castration are probably due to the absence of a hormone which is normally elaborated by the interstitial cells.

Further confirmation of the views just advanced is obtained from X-ray studies. These rays have a selective action in the sense that they destroy the cells of the testes, yet not those belonging to the interstitial gland. If the testicles of a roebuck are exposed to X-rays, the antlers do not undergo any alteration; here the generative cells proper have been destroyed. When the animal is castrated, which involves removal both of the external and internal glands, it loses its antlers. Clearly then the result of castration must be attributed to the loss of the interstitial hormone.

From the evidence presented we may conclude that the hormone or hormones that give rise to the male characteristics are due to the interstitial gland. The hormone is elaborated by the gland and passes directly into the blood stream. We are here therefore dealing with a gland belonging to the "ductless" group.

Professor Biedl says that "it is highly probable that, by the agency of its secretory product, the interstitial gland is responsible for the development of the male sexual gland from the indifferent genital trace. That it has a determining influence upon the normal development and maturity of the generative portion of the sexual gland, upon the formation of the secondary genital organs, and upon the existence and persistence of those morphological and biological characters which are the property of the male sex, is undoubted."

Our information regarding the interstitial cells of the female is meager, but whatever we do know points to their presence, and to their possessing a function similar to the cells of the male.

Grafting.—Brown-Séquard's suggestion that the active genital glands impart vigor to the body, due to the elaboration of an internal secretion, and the further fact that with old age the interstitial cells become fewer and smaller, must have suggested to physiologists a possible connection between old age and the interstitial glands. To test out such a possibility has led to a number of extremely interesting experiments. Research workers following Brown-Séquard showed that the addition to the food, or the injection into the blood, of testicular extracts, failed to change castrated into normal animals, or old animals into young ones. The reason for this is not easy to give, unless we assume that a chemical process involving an extraction of the hormone from the interstitial gland—an operation which is usually carried out by triturating the gland with sand and water, filtering and using the filtrate—causes an alteration of the hormone.

Dr. Voronoff remarks that "the injections of testicular juice have not had the result which Brown-Séquard expected from them, because the glandular extracts undergo rapid changes, do not contain the whole of the product of the internal secretion, and are even, at times, toxic." [1] This is obviously not true of some of the other ductless glands, notably the thyroid.

Next came the suggestion that a much better method than that of injection of gland would be the

[1] See *Life* by Serge Voronoff. E. P. Dutton & Company, New York.

grafting of a new tissue to take the place of the old. This held out hope, for many experiments with glands of the ductless variety pointed to the fact that they could be removed from their original positions in the body and grafted on to some distant tissue of the body, with no ill-effects, provided the graft "took"; that is, provided a healing condition set in whereby the blood vessels of the gland and those of the rest of the body would connect up, so as to allow the hormone of the ductless gland to enter the general circulation.

Upon ideas such as these are based the much-talked-of experiments of a Steinach and a Voronoff. "The grafting of a young sex gland in full activity," writes Dr. Voronoff, "means incorporating in the organism the very source of our organic action. Thus the body would be supplied not with a dead product, incomplete, often changed, introduced from time to time by means of subcutaneous injections, but a living organ carrying out its functions itself. To graft this gland is to place it in direct communication with our blood vessels, which will undertake to transport the precious fluid in proportion to its formation in the intimacy of our tissues."

Before we proceed to these experiments a word must be said as to the methods used in the transplanation of tissues. There are two ways of doing this: either by merely inserting a small strip of tissue in the desired place, or by first carefully connecting the engrafted with the main tissue by stitching suitable blood vessels together. In the first case general circulation in the engrafted tissue is for a time delayed; in the second, it is immediately set up. The first method has been extensively used in replacing small portions of skin in burns. It has even been employed, and with success, to interchange the ovaries in hens, and to replace the thyroid and parathyroid. The second method, more difficult to execute, has the advantage in that the transplanted tissue may be of much larger size. By this method the kidney, the spleen, and even a limb have been transferred from one dog to another; and segments of arteries that have been kept in cold storage, or preserved in formaldehyde, have successfully replaced portions that had been removed from their positions in the body.

It should be pointed out that, as a rule, transplantation is successful only if tissues of the *same* species are used, though experiments have shown that the arteries of a dog can be transplanted to a cat; and that further, much of the work recorded has been done on animals rather than on human beings. In the hands of Carrel and other famous surgeons active in the late war, the art of grafting has made long strides. The French doctors have proved themselves particularly skillful in the art.

The famous Siamese twins illustrate a perfect

graft for which Nature is responsible. Here the circulatory system resembled that of one rather than of two beings. An imitation of this kind has been produced in white rats by opening up the membrane enclosing the abdomen and stitching the skin and connective tissues together.

Steinach's contribution.—In approaching the work of Steinach and others we must be more than ever careful to distinguish fact from fancy; to relate actual experiments without drawing fanciful conclusions from them. Professor Steinach, one of the best known among the workers on the functions of the sex glands, is a member of the Vienna Academy of Sciences, and Director of its Biological Institute. His work as a biologist has commanded the respect and admiration of his colleagues throughout the world, though some of his interpretations of his work have not gone unchallenged.

These prefatory remarks are necessary because the impression seems to have gained ground that he is one of many quacks that fatten on the public's credulity. Such, in fact, was the impression gained by many newspaper readers after hearing of the sudden death of a rich Australian, who was to deliver a public lecture in London on "How I was made twenty years younger by the method of Dr. Steinach," and who died under somewhat mysterious circumstances on the very eve of the lecture. The newspaper drew the moral of St. Mark that new wine should not be poured into old bottles!

Prior to 1910, when Dr. Steinach published the first of his experiments, it had already been known that the generative apparatus consists of an internal and external secretion, and there were even then indications that the hormone, or hormones, produced by the internal secretion (in the interstitial cells) controlled sex characteristics. Professor Steinach wished first to convince himself that what was generally believed had foundation in fact: that the internal secretion actually controlled sex characters. With this in view he castrated a number of *male* rats and guinea-pigs and planted *ovaries* under the skin. The animals developed marked feminine features: their breasts and nipples increased in size and they were sought by the males. Previous experiments on the transplantation of the testes had shown that though the reproductive glands proper decayed and the interstitial glands did not, yet the sex desire remained. These experiments indicated two things: first, that the generative glands have much to do with determining the sex of the animal; secondly, that the portion of the glands particularly responsible is the interstitial gland.

Old age.—As already pointed out, the work of Brown-Séquard suggested a connection between the decay of the genital glands and the onset of old age. This was the next problem that Steinach set himself to elucidate. His idea was to transplant the sexual glands of young rats to the system of old ones—to put "new wine into old bottles." Much preliminary work was necessary in order to determine the conditions for developing healthy breeds of rats. The extensive vitamine studies, in which rats are very largely employed, had then hardly begun; otherwise such difficulties could have been avoided. Steinach succeeded not only in improving the general condition of the old rats, not only in renewing the sex instinct, but actually in prolonging the life of the animal. "Three weeks after the operation the whole behavior changed: the rat (an old one) began to keep its head erect, came forward from its hole, paid a lively interest toward the outer world, and its hair began to grow smooth and shiny. When another male rat had been let into its cage it attacked him immediately, and when a female rat was let in, it eagerly performed the sexual function, and as only one *vas deferens* (see subsequent paragraph) was ligated, the coitus was successful. The young of the female fecundated by the rejuvenated rats thrived well, and propagated further."

Our discussion of the interstitial glands has already shown that in the investigations of this ductless gland, one of three methods in operating technique may be employed: the duct from the testicle connecting the testis with the ejaculatory tube (the *vas deferens*) is cut off, thereby stopping the production of the external secretion, and hence the production of the reproductive elements; or the genital glands are submitted to the effects of X-rays, causing the less resistive reproductive elements to be destroyed. In either case the interstitial cells are not only left intact, but actually multiply and produce hormone in quantity (and quality?) sufficient to rejuvenate (?) the organism. There is still a third method, and that is to transplant the gland from a vigorous animal to the one under experimentation. We have already pointed out that the transplantation of testicles has shown that under such conditions, the reproductive elements disappear, but the interstitial cells continue to thrive. Here again the organism is favorably affected.

All three methods point to the interstitial gland as the source from which rejuvenation springs.

With Steinach the question now arose as to which of the three methods would be the one best adapted for his purpose. In his experiments with human beings he decided upon the first one—the ligature of the *vas deferens*—as being the simplest to perform, and the one least likely to give rise to unfavorable post-operative symptoms. "By stimulating the action of the interstitial gland at the expense of the generative function, it is Steinach's idea to bring about a rejuvenation process in older people by the resuscitation and renewal of the weakening secondary characters." We shall see that Dr. Voronoff selected the more difficult grafting procedure.

"The operation is a very simple one," writes Dr. Steinach; "absolutely painless. Quite free of any risk whatever. Takes no longer than fifteen minutes to perform. Seven to ten days in the hospital are all that's needed afterwards. But the operation must be performed with minute precision; and for that reason I cannot guarantee results unless I am personally present."

With the help of a local anesthetic a small incision is made and the *vas deferens* tied and cut off. Care is taken not to injure the blood vessels near by. The result is that the seminal vesicle (either one of the two reservoirs for the semen) and the interstitial gland are completely cut off from one another; and this in turn gives rise to a multiplication of the interstitial cells, and to an increase of the hormone produced by them.

Professor Steinach has performed the operation on men on several occasions. In some instances these men were fairly young but physically weak; in others, the subjects were senile men. "The appearance of the subjects became youngish, fresh, their bodily strength increased, the tremor of their hand disappeared, memory and will power returned, and the sexual power was restored."

Dr. Steinach has not had the same success with females as with males. With females the X-ray method holds out the most hope, but even with this method the results have not been very encouraging. Just why this should be so is not clear.

Of the cases described, one may be quoted as typical. "He was 71 years of age, controller of a large business. The man was actually ill, and showed the usual signs of old age and decay, such as dizziness, poor breathing, heart weakness, shakiness and intense fatigue. For eight years ambition had been practically non-existent. Some months after the operation there was remarkable evidence that the conditions of senility had been checked. The man's power came back. It is still increasing. We got a letter from him recently (Dec., 1919) in which he told us of the extraordinary change that had taken place in his condition. His appetite had come back. His old-time spells of depression had given way to a new joy in life. He had become fresh in looks and elastic in body. His dizziness had gone. His hand had become firm again. He specially mentioned the fact that he was again able to think clearly, as in earlier years, and that his memory had greatly improved. Also, that whereas in earlier years he had needed the barber once only in two or three weeks, he now visited his beard-trimmer once a week."

Dr. Steinach has recently summarized his work in an article for a technical journal entitled *Verjüngung durch experimentelle Neubelebung der alternden Pubertätsdrüse;* which may be translated into "Rejuvenation through experimental revivifying of the senescent puberty (interstitial) glands." This article under the short title of *Verjüngung* has since been published in book form (by Springer, of Berlin).

Professor Steinach's work receives support from that of Dr. Lichtenstern, a lecturer at the University of Vienna, who has successfully performed 21 transplantations and 36 operations in cases of senility. Yet we must also observe that Professor Kolber, another colleague of Dr. Steinach's, who has made a careful histological study of the interstitial cells with the view to elucidating their function, is not wholly in agreement with Steinach; and perhaps in the form of a word of warning, if only to ward off too much optimism, it may well be asked why the repairment of a part of the organism—a very important part, we admit—that has become very generally dilapidated, should rejuvenate the entire bodily machine. It must be remembered that not only the interstitial gland, but the other ductless glands, are affected with the onset of old age,— to say nothing of the rest of the organism.

Voronoff.—Dr. Serge Voronoff is another investigator who has lately appeared before newspaper readers. He is the Director of Experimental Surgery at the Laboratory of Physiology of the Collège de France, a position once held by perhaps the greatest of all physiologists, Claude Bernard.

Unlike Steinach, Voronoff decided on the grafting procedure. His preliminary experiments on animals—on sheep and goats—are valuable contributions to the literature of the subject. His subsequent work on man was limited to two recorded cases, and here the graft does not refer to the sexual glands at all, but to the thyroid. We are, however, assured that but for the scarcity of orang-utangs more cases could have been presented, and the effect of the transplantation of the sex gland of the monkey to man could also have been recorded. But let us proceed to those portions of Voronoff's work that are of value to us.

Dr. Voronoff tells us that he has made some 120 different experiments on the effects of testicular grafting in sheep and goats. Grafts were attempted on normal males and females, castrated males and females, and old males incapable of reproduction. As the results with the females were in no instance encouraging, these may be dismissed altogether, and our attention confined to the males.

Of the two methods of grafting which are employed, and which have already been described, Dr. Voronoff chose the simpler, namely, the transplantation under the skin or in the peritoneum.

Later on he found that the very best results were obtained when the grafts were placed in the vaginal tunics. As this simple method of grafting gives better results when fragments rather than whole organs are used—in other words, when smaller rather than larger pieces are used—Voronoff grafted fragments obtained from a whole testicle.

In one set of experiments three he-goats were castrated and testicular grafts introduced. One at the end of four, another at the end of twelve, and a third at the end of sixteen months all showed "magnificent horns such as are never seen on castrated animals." They were lively, belligerent, did not grow fat, and showed much sex ardor.

In another experiment a ram 12 to 14 years old,— corresponding to about 80 years in man—which could hardly totter, had implanted the fragments of a testicle obtained from a young ram. "Two months after the graft had been effected the animal was completely transformed. His urinal incontinence had disappeared; so had tremblings of the legs; and he no longer looked afraid. His bodily carriage had become magnificent, he behaved in a lively and aggressive manner. The old ram had taken on the appearance of remarkable youth and vigor. He was isolated in a small stable, together with a young ewe-lamb, which afforded the opportunity for observing not only the awakening of his sexual instincts, which he had lost years ago, but also the following more tangible result: the ewe-lamb covered by him in September, 1918, dropped a vigorous lamb in February, 1919. There is nothing in the fact to cause surprise. Old animals, like very aged men, occasionally still possess spermatozoids which are altogether alive, but it is the atrophy of the internal secretive glands which prevents their experiencing the sexual appetite and manifesting their virility."

Dr. Voronoff's next procedure with the rejuvenated animal was excellent from the scientific standpoint. He removed the graft. Three months later the animal had completely aged. Then he reimplanted another graft from the testicle of a younger animal. Once again the animal showed signs of rejuvenation. Nothing in the whole book approaches in value this particular experiment.

At the time of writing the graft had been in its place for a year and the animal was still in excellent condition. Even if we could reason by analogy, —which in this instance we should be careful not to do—we could hardly draw any far-reaching conclusions as to the length of time old age could be warded off in man.

Having successfully performed experiments on animals, Dr. Voronoff next turned his attention to man. Here a difficulty immediately arose. How was he to get a healthy young man's gland in order to implant it into an old man's system? If only the foolish public would permit him to preserve the glands of all healthy young men that are accidentally killed! But the foolish public will do nothing of the sort, and experience has shown that only glands of the same species are of any value. There is one way out of the difficulty. It is to make use of the higher form of ape, such as the orang-utang or chimpanzee, whose near relationship to man is to-day firmly established.

Why in the only two recorded cases of grafts of monkey glands on human beings, men suffering from a deficiency of the thyroid hormone were selected is not clear. It is possible that no aged men willing to be experimented on with testicular extracts were just then on hand; neither is it easy to obtain monkeys whenever you need them.

"There is no question," writes Dr. Nagel, "that Voronoff had a great idea and that operations according to his theory were temporarily encouraging. They do not, however, last. As soon as the gland that has been transplanted has been absorbed by the body, the subject of the experiment returns to his original condition.

"In Steinach's operation this is not the case. He does not transplant. He merely ties off the offending gland, the inert part is absorbed by the body, and a new one is grown by nature in place of the old, useless one.

"What does a gardener do when he wants to invigorate a tree? He cuts off the dead and dying branches and immediately new ones are thrown out by nature.

"An illustration may be had in common surgery. When it becomes necessary to sever an artery the surgeon ties it off. Very shortly the checked circulation would cause serious trouble. But nature takes care of that. The moment the artery is tied off little shoots spring out from the walls of both ends of the main artery, growing rapidly until they unite around the useless part and the artery is whole again.

"It is just that way with the gland tied off by Steinach. A new one grows, and, being new, is far better than any attempted rejuvenation of the old, dying one."

We may summarize the work both of Professor Steinach and of Dr. Voronoff by saying that while experiments on animals have yielded results that are of great value and that are extremely suggestive, those carried out on human beings have been altogether too few to warrant any hasty generalizations. We are far, perhaps very far, from the time when the mere grafting of one of the ductless glands will cause old age with all its concomitant horrors to disappear.

Before dismissing this subject it should be mentioned that Lorand, a Spanish investigator, is of the opinion that senility is the result of primary degeneration of *all* the ductless glands, and not merely the genital glands. In one of his latest papers (June, 1921) he says that he has been very successful in arresting the signs of senility by a combination of thyroid and genital gland extract treatment, especially in women. But let not the enthusiasm of a few individuals make us too credulous. The subject has pitfalls at every step, and not only the innocent layman, but many a physician is apt to slip.

Pubertas precox.—So far our discussion has been limited to cases of diminished or hypo-secretion of the interstitial gland. We have seen that such cases may occur naturally, as in eunuchoids; and we are probably not far wrong in including old men and perhaps old women under this heading, though their condition is much more complex. We have already seen how the *hypo* condition can be deliberately brought about by means of castration. The grafting of another gland from a more vigorous animal of the same species, or the adoption of an operative technique whereby the interstitial cells, and consequently their hormonic output, increases, brings about cures that may, or may not be permanent (our information on this point is nowhere near as complete as it might be). Now the question naturally arises, are instances of excessive or hyper-secretion met with? And is it possible to induce such a hyper-secretion just as, let us say, it is possible to induce a hyper-thyroidism?

In answering the second question first, we may point out that the number of experiments dealing with induced hyper-secretion of the interstitial gland is limited. Those of Brown-Séquard on the injection of testicular extracts have already been described, and we know that his experiments left much to be desired. There are, however, individuals that have been known to suffer from an excessive secretion. As might be anticipated, this occurs, if at all, at a very early age. Such cases go under the name of *pubertas precox,* to indicate early sexual maturity. They are often associated with tumor growths.

One of the most interesting examples of hypersecretion is that given by Sacchi. A boy had developed quite normally until he was five and a half years old. From that time on until he was nine and a half the boy changed remarkably. His voice deepened, hair grew on his face, and when he had reached the age of nine and a half years he actually had a black beard. An examination of his left testicle revealed a tumor growth. This was removed. The child became quite normal again.

Another example is that of a girl of seven, who had already menstruated a number of times, and who had well-developed breasts and pubic hair. A tumor of the ovary was removed, and the girl grew up into normal womanhood.

CHAPTER VII

THE PANCREAS AND THE LIVER

Along the lower part of the stomach, and connected with the spleen, is a gland shaped like a dog's tongue, weighing about three and a half ounces, and having a length of about eight inches. This gland, the "pancreas"—from the Greek "all flesh"—is popularly known as "sweetbread."

The pancreas.—The pancreas is the most important organ of digestion. Its glands secrete substances (enzymes) which, by means of a duct, find their way into the small intestine, and there attack the food coming from the stomach, breaking the food particles up into such simple chemical substances as to make them fit for absorption by the blood and lymph.

It was not known—definitely at least—until 1889 that this organ had any other function than that of taking part in the digestive process. Until then the pancreas was regarded as a gland producing an *external* secretion only—that is, a secretion which, once manufactured, is sent by means of a tube to the surface, and, in this particular case, ultimately finds its way to the small intestine. In 1889 Minkowski and von Mering, two German investigators, proved that in addition to its external secretion, it also produces an internal one—a secretion that passes directly into the blood stream, and one that regulates the sugar metabolism of the body. Since this sugar or carbohydrate metabolism is one that is inseparably connected with an important function of the liver, a few words must be said regarding that organ.

The liver.—The liver, which is situated on the right side of the body and partly covers the stomach, is the largest gland of the body. It weighs from three and a half to four pounds. It produces an external secretion, the bile, which is either first stored in the gall bladder until wanted, or is sent directly into the small intestine, meeting the secretion from the pancreas, and helping the latter in its work of digestion. But it does far more. It is the seat of carbohydrate metabolism in the body.

Carbohydrate metabolism.—Now what do I mean by "carbohydrate metabolism"? Carbohydrates are one of the three classes of foodstuffs, and they include such substances as sugar and starch (in flour, say). When we eat one or more of these carbohydrates they may or may not be broken up in the digestive tube into chemically simpler substances. If they are chemically complex, such as starch, they will be; if they are chemically simple, such as glucose, they will not be. In any case, enzymes secreted by the salivary, pancreatic and intestinal glands and sent into the digestive tract, are ever ready to convert complex into simple carbohydrates. The chemically simple carbohydrates, of which there are three well-known types (glucose, levulose, galactose), are absorbed by the blood, sent to the liver and there stored in the form of glycogen or "animal starch."

Glycogen.—Chemically, glycogen is about as complex as starch. If, therefore, the simple carbohydrates, obtained from starch and similar bodies by the breaking up of the starch molecule, are converted in the liver into glycogen, it can mean but one thing: that there must be a recombination of the simple carbohydrate molecules to form glycogen.

But this is not all. Whenever the body needs carbohydrate—which it does much of the time for its energy supply—it calls on this glycogen stored in the liver. The glycogen is first converted back again into a simple sugar—but one only—namely glucose, and this glucose is sent via the blood to the various muscles, where it is "burnt" to supply the fuel needs of the body.

Summarizing then the glycogen function of the liver, we may say that the carbohydrates we eat, after an appropriate preliminary simplification in the digestive tract, are transferred to the liver by the blood, and stored there in the form of glycogen. Whenever the body needs carbohydrates for its various energy purposes, the glycogen is decomposed into glucose, and this substance is sent to the various tissues where it is "burnt" or "oxidized" into carbon dioxide and water.

The liver a ductless gland.—You will remember that we defined a ductless gland as one producing a secretion which passed directly into the blood, and which secretion contained a something (hormone) that influenced other organs to various modes of activity.   Can the liver be viewed as a ductless gland?   It certainly produces an internal secretion in the sense of being able to take material (simple carbohydrates) from the blood, converting it into other products (first into glycogen and then into glucose), and passing the glucose directly into the blood.   But where is the hormone, the substance that influences the activity of other organs?   We must look to the pancreas for such a substance.

Claude Bernard.—At any rate, the glycogen function of the liver [1] is of extraordinary interest,

---

[1] Not only does the liver manufacture glycogen, but it also makes urea from ammonia and carbon dioxide.   This urea is carried away by the blood and finds its way into the urine.   80 per cent. of the total nitrogenous products (representing substances derived from protein) in the urine is in the form of urea. The formation of urea, like that of glycogen, is evidence of the internal secretory mechanism of the liver.   This by no means exhausts the functions of the liver, a veritable storehouse of wonders.

It should be remarked at this point that a rigid definition of since its discovery and study led the celebrated French physiologist, Claude Bernard, to formulate for the very first time the conception of an internal secretory gland. In his justly famous *Leçons de physiologie expérimentale,* published as far back as 1855, he writes: "For a time a false conception has been current as to what a secretory organ consists of. It was believed that all secretions must be poured upon an internal or external surface, and that all secretory organs must necessarily be provided with an excretory duct for the purpose of conveying to the exterior the products of secretion. The case of the liver establishes in a most lucid manner that there are internal secretions, that is, secretions which, instead of being carried to the exterior, are diffused directly into the blood. . . . It is now firmly established that the liver has two functions of the nature of secretion. The first,

an internal secretory gland would rule out the liver, since neither glucose nor glycogen, the substances manufactured by it, can be regarded as hormones. However, it has historical importance in connection with Claude Bernard's work.

If we want to stretch our definition, we could well afford to include such a substance as carbon dioxide in a list of hormones. It is the amount of this gas in the blood, and not, as has been supposed, the amount of oxygen, that controls the process of breathing. Increased exertion, as in running, causes increased production of carbon dioxide, and this in turn stimulates more rapid breathing.

A very practical use of this new knowledge has been made by Yandel Henderson, the Yale physiologist. It is only a too well known fact that after anesthesia the patient's breathing is poor. Professor Henderson shows that this can be improved by *adding carbon dioxide* to the air that is administered.

the external secretion, produces the bile, which flows to the exterior; the second, the internal secretion, forms sugar which immediately enters into the blood of the general circulation."

Claude Bernard, in fact, coined the phrase "internal secretion," though his views as to the influence of this secretion on the rest of the body, if not vague, were misleading.

The pancreas and its internal secretion.—It is now time to turn to Minkowski and von Mering's experiments which proved that the pancreas regulates the sugar metabolism by means of a hormone which it develops. These investigators found that by carefully and completely extirpating the pancreas in a dog, the animal develops diabetes. Such an operation was invariably followed by the death of the animal within a few weeks, but during those weeks it suffered severely from the sugar disease.

Diabetes, as you may know, is the—unfortunately—all too common disease wherein the body cannot use or "burn" all the sugar that it gets, with the result that we find it present in excessive quantity in the blood (hyperglycemia), and sooner or later an excessive quantity of sugar appears in the urine (glycosuria). Since the liver is the central organ for sugar metabolism, we must look for a disturbance in the liver function. The German investigators showed that such a disturbance occurs when the pancreas is cut out of the system.

But you may ask, how do we know that the pancreas develops an internal secretion the hormone of which travels through the blood to the liver? Why cannot we assume that whatever influence the pancreas exerts on the liver is due to its external secretion—to the pancreatic juice which, by means of a duct, is poured into the small intestine? We can answer these questions very definitely. It has been shown that incomplete removal of the pancreas—as little as one-fourth to one-fifth of the total need be left—prevents diabetes, even though all connections with the small intestine have been interrupted, and all flow of pancreatic juice has stopped. Or again, the ducts of the gland may be tied, or, what amounts to the same thing, filled with paraffin, and no glycosuria results. But the most convincing proof that the pancreas develops an internal secretion is this: after complete extirpation of the gland, and after the onset of the disease, the symptoms of diabetes disappear upon the grafting of a piece of pancreas under the skin or into the abdominal cavity. Here whatever communication is open between the grafted pancreas and the other organs of the body is by means of the blood and the blood only.

Recent experimentation tends to show that the internal secretion of the pancreas is not developed by the pancreatic cells themselves but by peculiarly shaped groups of cells, scattered throughout the pancreas and known collectively as the "Islands of Langerhans," after their discoverer. Ssobelow, a Russian, tied the pancreatic duct and noticed that the pancreatic cells decayed; but not so the cells of the "Islands of Langerhans." We have already noticed that such an operation is not followed by the appearance of sugar in the urine, whereas sugar does appear upon the complete extirpation of the pancreas. Hence it seems logical to conclude that the pancreatic hormone is produced by the "Islands" rather than by the pancreatic cells proper. This view finds corroboration in the observations of clinicians who have noticed that in cases of diabetes the "Islands of Langerhans" show degeneration.

How the amount of sugar in the blood is regulated.—Just how the pancreatic hormone regulates the sugar output is problematical. Some claim that it takes a hand in the oxidation of glucose to carbon dioxide and water; and that in its absence the glucose cannot be so oxidized, whence it accumulates in the blood and finds its way into the urine. Others, on the contrary, are of the opinion that it has nothing at all to do with the oxidation of glucose, but that it does regulate the amount of glucose to be formed from the glycogen in the liver. In the absence of the pancreatic hormone, the regulatory influence being absent, excessive quantities of glycogen are converted into glucose, which finds its way into the blood and then into the urine. An attractive hypothesis of this type claims that there really are two opposing forces that work in the normal man,—an accelerating force, due to the hormone from the adrenal glands, which accelerates the conversion of glycogen into glucose, and a retarding force, due to the pancreatic hormone, which tends to slow up and regulate such a reaction. To support this hypothesis, Dr. Zuelzer, its author, describes an experiment with a dog that had had its pancreas removed and its adrenal veins ligated. No diabetes followed.

We may represent what happens as follows: carbohydrates $\longrightarrow$ stored in liver as glycogen $\longrightarrow$ given out as glucose. The blood, when analyzed is always found to contain a small amount of glucose. Under normal conditions that glucose is fairly constant in amount, ranging from 0.07 to 0.1 per cent. Even in a starving animal the blood sugar is found to remain constant. This makes us believe that under such conditions, with all carbohydrate virtually absent from the body, part of the fat molecule, and perhaps part of the protein molecule, is converted into sugar. Only in pathological cases, such as are met with in diabetes, does the amount of blood sugar materially increase.

Now the question arises, by what mechanism is the blood sugar of the normal human being kept constant? Why despite the varying quantity of carbohydrate eaten from day to day does not the blood sugar fluctuate correspondingly? A number of hypotheses have been advanced to explain this controlling mechanism. Very recently (April, 1921) Dr. Langfeldt has proposed a theory which includes the best elements of the theories of other investigators as well as a little of his own speculation. In essence it is this: It is evident that there must be at least two controlling factors, one involving the conversion of carbohydrates into glycogen, and the other the conversion of glycogen into glucose. The first part of the reaction, the glycogen synthesis, is controlled by the pancreatic hormone; the second, the breaking down of glycogen to glucose, with the object of meeting the energy requirements of the muscles, is a more complex affair depending for one thing upon the degree of acidity of the blood. It has been shown that the glycogenase, the enzyme responsible for the transformation of glycogen to dextrose, is most active when the blood is slightly acid. Ingenious physical chemists have devised quantitative methods for determining the exact state of acidity not only of blood but of any liquid. If then the acidity of the blood is below or above a certain optimum, the glycogenase is not so active, and less glucose is formed —a rare and abnormal condition.

On the other hand, a more common occurrence is where the pancreatic hormone fails to function, due to the failure to function of the pancreas; in that case little of the carbohydrates is converted to glycogen, and the excess sugar floods the blood and the urine. We here have a typical case of diabetes.

The "Allen treatment" for diabetes.—No mention of diabetes is possible without referring to the *Allen* treatment of this disease. Dr. Allen, until recently with the Rockefeller Institute, and now head of the Physiatric Institute, Morristown, N.J., has had such remarkably good results with his "fasting treatments," that physicians all over the world have adopted his method, and there can be little room for doubt as to its success. Only recently (June, 1921) at the Wiesbaden Congress of Internal Medicine, the two foremost authorities on diabetes in Europe, von Noorden and Minkowski, sang its praises.

The treatment is as follows:[1] A preliminary fast is taken until the urine is free from sugar. This will usually take less than four days. During that time water is allowed and also, to a certain extent, tea and coffee. Following the fast, carbohydrate food is gradually added, at first in the form of green vegetables. Coincident with the addition of carbohydrate, or in place of it if the carbohydrate tolerance is very low, protein is added to the

[1] This is described in the *Handbook of Therapy*, published by the American Medical Association.

diet in small but gradually increasing amounts until glycosuria occurs. Fats are added in small amounts during the time of addition of carbohydrates and proteins. Frequent urine examinations are made, either by the medical attendant or by the patient himself [1] and the diet regulated accordingly.

[1] Some Fehling's-Benedict solution, which is blue in color, is heated in a test tube until it boils, and an equal volume of the urine under examination is added. The mixture is heated for a minute or two. If sugar is present, a red precipitate will make its appearance, and the amount of precipitate will give some rough idea of the amount of sugar present. A modification of this method lends itself to a fairly exact quantitative estimation.

The more recent work on the subject has centered itself as much in determining the quantity of sugar in the blood as in that of the urine, and much ingenuity has been expended to devise methods that are refined enough when dealing with quantities of blood even as small as a drop or two—the amount obtained from a prick in a finger. The blood being not only the medium by which materials are taken in by the body, but also that by which they are given out, it need cause little surprise that blood analyses should throw much light on what takes place in the body.

CHAPTER VIII

THE INTESTINAL HORMONE

We have already referred (page 4) to the work of the English physiologists, Bayliss and Starling, on secretin. This work is of such a fundamental character, that it well deserves a chapter all to itself.

In the preceding chapter we have shown how the glycogenic function is controlled by a hormone developed by the pancreas. We shall now proceed to show that the pancreatic juice in turn is controlled by a hormone which has its origin in the lining of the upper part of the small intestine.

Digestion in the small intestine.—Even at the risk of repeating what was said in the introduction, let us state what happens in the small intestine during digestion. You will remember that when the valve connecting the stomach and intestine is opened, the food passes from the former into the latter. Here the food is acted upon by three liquids, each containing either enzymes or other substances that help to simplify the food to the point where it can be absorbed by the blood and lymph and sent to the cells. One of these liquids is elaborated by the intestine itself; another, the bile, comes from the liver; and the third, the pancreatic juice, from the pancreas.

The last two reach the intestine by means of ducts or tubes. They represent typical "external secretions," in contradistinction to the "internal secretions" that we have studied, and that flow directly into the blood.

The hormone in the intestine.—Now the question arises, why whenever food enters the intestine, and only then, do bile and pancreatic juice also begin to flow into it? The easiest answer, the most obvious one, is that there is a nervous mechanism involved; that the brain correlates the activities of these organs. Such a theory had a distinguished supporter in the person of Pavlov, the Russian physiologist, whose present plight has been so graphically described by H. G. Wells, the novelist who is also a scientist. We are about to show that Pavlov's theory of brain interference is untenable; that all nerve connections between the brain and intestine can be severed without stopping the flow of pancreatic juice; that the chief factor that brings about this coördination—at least in so far as the flow of pancreatic juice is concerned, and, to a less extent, the bile,—is a hormone elaborated by the small intestine; that this hormone is produced whenever the acid food from the stomach finds its way into the intestine; and that this hormone, once produced, finds its way to the pancreas via the blood stream. You will notice incidentally that if what we now say is true, the intestine is similar to the pancreas and, according to Claude Bernard, to the liver, in that it elaborates both an internal and an external secretion.

The proof that the flow of pancreatic juice is controlled by a hormone elaborated by the intestine is due to Bayliss and Starling, the presiding geniuses of the department of physiology at University College, London. Their experiment marks a milestone in the development of our knowledge of internal secretions; and we shall offer no excuse for quoting from part of their celebrated paper on the subject, technical though it is.

Bayliss and Starling's experiment.—"On January 16, 1902, a bitch of about 6 kilos (13-14 pounds) weight which had been fed about 18 hours previously, was given a hypodermic injection of morphia some three hours before the experiment. The nervous masses and celiac (pertaining to the abdomen) axis were completely removed, and both vagi (nerves situated in this part of the body) cut. A loop of jejenum (a portion of the small intestine) was tied at both ends, and the nerves supplying it were carefully dissected out and divided, so that the piece of intestine was connected to the body of the animal merely by its arteries and veins. A cannula (tube for inserting into body) was inserted in the large pancreatic duct and the drops of secretion recorded. The blood pressure in the carotid (the principal artery in the neck) was also recorded. The animal was in a warm saline bath, and under artificial respiration.

"The introduction of about 20 cubic centimeters (about 1-25 of a pint) or about one tablespoon of four-tenths per cent. of hydrochloric acid [1] into the duodenum (the first portion of the small intestine) produced a well-marked secretion of one drop every 20 seconds, lasting for some six minutes; this result merely confirms previous work. But—and this is the important part of the experiment, and the turning point of the whole research—the introduction of 10 cubic centimeters of the same acid into the enervated loop of jejenum produced a similar and equally marked effect.

"Now, since this part of the intestine was completely cut off from nervous connection with the pancreas, the conclusion was inevitable that the effect was produced by some chemical substance finding its way into the loop of jejenum in question, and being carried in the blood stream to the pancreatic cells. Wertheimer and Le Page have shown,

[1] Hydrochloric acid is found in the stomach under normal conditions. It is approximately four-tenths of one per cent. in strength. Of course the object in introducing the acid is to prove that it, and it alone, coming from the stomach, liberates the hormone in the intestine.

however, that acid alone introduced into the circulation has no effect on the pancreatic secretion, so that the body of which we were in search could not be the acid itself."

This suggested that there may be a *something* in the wall of the intestine which was responsible for the action. "The next step in our experiment was plain—namely, to cut out the loop of jejenum, scrape off the mucous membrane (tissue covering the surface), rub it up with sand and four-tenths per cent. hydrochloric acid in a mortar, filter through cotton wool to get rid of lumps and sand, and inject the extract into a vein. . . . After a period of about 20 seconds, we obtained a flow of pancreatic juice at more than twice the rate produced at the beginning of the experiment by introduction of acid into the duodenum."

A pretty variation of this experiment was performed by Enriquez and Hallion. They conveyed the blood stream from the vessels of one dog (A) into those of another (B) and found that after injecting acid into the small intestine of dog (A), pancreatic juice began to flow in dog (B)!

Secretin.—Though the hormone responsible for the flow of pancreatic juice has not been isolated in the pure state, its discoverers have given it the name "secretin" (from the Greek "to excite.") Since a watery extract of the mucous membrane of the intestine when injected into the blood stream, fails to cause a flow of pancreatic juice, and since hydrochloric acid alone is no better, but since when the two are mixed we do get a response, Professors Bayliss and Starling have advanced the hypothesis that in its original form the secretin is in an inactive state—the "pro-secretin" state they term it; and that the hormone becomes active only when the acid converts the pro-secretin into secretin. The function of the acid coming from the stomach, then, is to convert the inactive into an active hormone. The fact that solutions containing secretin can be boiled without destroying the hormone suggests that this hormone—and others?—are quite distinct from either vitamines or enzymes, both of which are quite susceptible to increases in temperature.

An objection to the work of Bayliss and Starling. —The Italian physiologist, Luciani, has criticized the work of the English scientists. He writes: [1] "Popielski and his pupils have recently published a series of experiments and conclusions which completely refute the *secretin theory.* Popielski states that the substance extracted after the extraction of the duodenal mucosa with hydrochloric acid is not specific, but may, on the contrary, be obtained by simple hydrolysis, from any glandular, muscular, or even nervous tissue. . . . But the following is the most cogent of Popielski's arguments.

[1] *Human Physiology,* Volume 2.

On repeating the injections of secretin many times in equal doses, he observed a conspicuous secretion after the first dose, less after the second, still less after the third, till the substance rapidly became ineffective. Now, the introduction of acid in the duodenum, however often repeated, invariably excites pancreatic secretion proportional to the quantity of acid introduced. The body evidently reacts to the introduction of secretin by forming an antibody capable of fixing it and annulling its action; this suggests that it is not a substance normally developed by the body, but is an artificial extraneous product."

Despite this criticism, the secretin theory has been very generally adopted; to every physiologist who cites an experience in opposition to it, there are twenty who cite experiments that support it.

Gastric secretin.—While on the subject of digestive juices, reference may be made to the gastric juice, a fluid manufactured in the walls of the stomach. Pavlov has conclusively shown that the flow of gastric juice is unquestionably controlled by the brain, since the severance of all nerve connections stops the flow. This would seem to show that the origin of the flow of gastric and pancreatic juices is fundamentally different. However, Dr. Edkins has been able to show that, in addition to a nervous reflex, there is also a chemical stimulus involved. His experiment followed the lines of Bayliss and

Starling. A piece of mucous membrane from the stomach, when extracted with acid, and the acid injected into the blood, caused a flow of gastric juice. Edkins called the hormone responsible for this action the *gastric secretin,* to distinguish it from the intestinal secretion.

Secretin and vitamine.—We have pointed out in this chapter that secretin withstands the temperature of boiling water, a fact speaking against its identity with any vitamine. However, Dr. Carl Voegtlin, the government chemist in the Hygienic Department at Washington, has recently performed experiments to show that secretin and vitamine B (the anti-neuritic vitamine—that is, the one that cures beriberi in man and polyneuritis in pigeons),[1] are similar in properties, if not actually one and the same substance. Towards chemical reagents, such as wood alcohol, silver, lead and barium salts, they behave alike. "Secretin preparations from the duodenum of hogs relieved to some extent the neuritic symptoms, and the anti-neuritic vitamine from brewer's yeast on injection into dogs stimulated the pancreatic and biliary secretions." Whether this be so or not we cannot say definitely; the work needs confirmation.

[1] See the chapter on Beriberi in the author's book on Vitamines.

CHAPTER IX

THYMUS, SPLEEN, PINEAL, MAMMARY GLAND AND KIDNEY

I have grouped in this chapter a number of organs whose exact position in ductless glandular classification is debatable. Some of the organs are not glandular in the histological sense. With others we are not altogether certain that they produce an internal secretion or, more correctly, a specific hormone. A substance like carbon dioxide is not produced by a gland, yet as a regulator of respiratory activity and of the respiratory center it ought, perhaps, to be discussed in this chapter. On the other hand, a compound like urea is the product of an organ of internal secretion; but it is not yet certain whether it has any direct influence on the kidney. This chapter, then, is full of uncertainties.

THYMUS

This organ is situated in the neck near the thyroid. It seems to be of particular importance in the early life of the individual, though the most recent work tends to the opinion that it functions throughout life.   After the second year of life it grows less in size.   Here are some figures: At birth, 13.26 grams (approximately 30 grams equal one ounce); between one and five, 33 grams; between six and ten, 26 grams; 11 to 15, 37; 16 to 20, 25; 56 to 65, 16; 66 to 75, 6.

Function of the thymus.—What the function of the thymus is is a matter of constant debate.   A number of experiments point to the fact that its activity is connected with that of the sex glands; that the thymus, for a time, checks the development of the reproductive organs.   For example, the removal of the thymus (in animals) is said to accelerate sexual development, though it delays growth; and castrated animals show an enlarged thymus.

Uhlenhuth, of the Rockefeller Institute, has sponsored the theory that the thymus secretes a tetany-producing substance (see the chapter on the parathyroid) which is neutralized by the parathyroids. Some claim that the thymus is the principal reserve organ for nucleoprotein, an important type of protein particularly abundant in the nuclei of cells. Still others deny that the thymus is an endocrine gland.   For example, Hopkins, in an exhaustive review of the subject, says: "The evidence in favor of such a theory (that the thymus is a ductless

gland) is circumstantial at best and very meager.
It is equally difficult to prove that the thymus does
not produce a secretion, but the burden of proof
is upon those who support the former theory."

A very remarkable experiment by Gudernatsch
must be cited here, though it should be noted at the
outset that another investigator, Swingle, chal-
lenges Gudernatsch's statements. The latter found
that by feeding tadpoles with thymus extract, their
growth could be accelerated to a remarkable degree, but that metamorphosis to the frog state was
delayed. We have already seen in the chapter on
the thyroid that thyroid extract behaves in an opposite manner, in that feeding tadpoles with such
an extract accelerates the transformation into the
frog, but retards growth. Are we to assume that
the thymus is connected, for a time at least, with
thyroid activity?

The general concensus of opinion is that the extirpation of the thymus does not necessarily result
fatally, though it does give rise to a disordered development of the skeleton, such as may be seen in
a rickety child. Sciplades, a Hungarian investigator, is of the opinion that osteomalacia, a disease
characterized by a softening of the bones, is brought
about by the absence of a functioning thymus. This
is based on experiments with young dogs whose
thymus had been completely extirpated. "The

changes produced in the bones coincided, histologically, with the changes characteristic of human osteomalacia."

Thymus extract does not cure the disease, though when injected into the blood it lowers the arterial pressure and accelerates the heart beat.

A disease called "mors thymica," sometimes "thymic asthma," connected with difficult breathing of infants, and which has a sudden fatal termination ("thymus-death,") is said to be due to the hypertrophy (enlargement of organ) of the thymus.

SPLEEN

The spleen, like the thymus, is not a glandular organ, and hence is often omitted in the treatment of ductless glands. But it seems to develop an internal secretion, in that a hormone from the spleen passes through the blood to the pancreas and "activates" the ferment (enzyme) that attacks protein— trypsin. This proof, if not convincing, is of a somewhat more positive nature than in the case of the thymus, where our only reason for supposing that it manufactures a hormone is the claim made by some that when the thymus is extirpated the skeleton does not develop properly.

The function of the spleen, like that of the thymus, is shrouded in much mystery. We have just said that it is possible that its hormone activates the pancreatic trypsin. It should, however, be added that Pavlov, the Russian physiologist, has proved that the juice elaborated in the small intestine also contains a hormone (or enzyme?) which activates the trypsin.

From the large quantity of iron (in "organic" combination) that the spleen contains, and from studies in anemia, investigators have concluded that it plays a part both in the formation and destruction of the red blood corpuscles, but this is by no means certain. Others regard it as playing an important part in immunity from the active phagocytosis (destruction of micro-organisms by cells such as the leucocytes, or white corpuscles of the blood). The very recent work by Inlow disproves the theory that the spleen regulates the digestive power of the stomach—a claim based on some experiments which are cited to prove that the removal of the spleen diminishes the activity of the pepsin, the enzyme in the stomach.

Curiously enough, the extirpation of the organ was practised by the ancients, in the belief that it improves the "wind" in runners. Extirpation is not attended with fatal, or even particularly bad results. It is practised in a disease called "splenic anemia," "characterized by progressive enlargement of the spleen, attacks of anemia, and a tendency to hemorrhages. . . ." Complete recovery follows the removal of this organ. This, of course, makes it quite evident that the spleen cannot compare in importance with other ductless glands, such as the adrenal or the pancreas, the absence of which causes death. This, however, does not mean that the spleen is of no importance.

Eddy in a recent review cites the following in support of the theory that the spleen produces an internal secretion: 1. Changes in erythrocytes after splenectomy (removal of spleen) ; 2. Modification of blood picture after hyperplasia (abnormal multiplication of the tissue elements) of the spleen, ameliorated in some cases at least by splenectomy; 3. Specific effects on the red blood corpuscles of injection of splenic extract. He acknowledges that we know nothing of the chemical nature of the hormone, but suggests that the chief function of the spleen is to remove from the circulation the disintegrated erythrocytes (red blood cells), and to build erythrocytes (by stimulating the erythrogenic, or blood corpuscle building power of the bone marrow).

PINEAL (*"Epiphysis"*)

This is an organ the size of a pea, situated at the base of the brain, behind and above the pituitary. Descartes considered it the seat of the soul! Like the thymus, its importance seems to be chiefly in the early stages of its existence—if we are to believe many of the authors who have busied themselves with the organ. Professor Biedl claims that in adults "the gland is a negligible factor." He arrives at this conclusion from extirpation experiments. He has not, however, settled the question of extirpation in young animals.

Foà, an Italian, has removed the pineal from roosters, with the result that the testes hypertrophied. Horrax, of Chicago, has practised pinealectomy on guinea pigs; he states that the development of the testes becomes accelerated. The feeding of the desiccated pineal body to rats has had no influence on their growth (Finney, Baltimore). Another and more important experiment where 27 mentally deficient children at the Vineland farm were fed with a pineal extract, led to no noticeable effect.

We do know what pathological growths of the pineal gland in children will give rise to. "In the 70 cases on record of tumor of the pineal gland, most were in adults, but ten were in boys below the age of puberty; and these all presented precocious and pronounced development of the primary and secondary sexual characteristics, and some a certain degree of mental precocity." (Zandren); which points to the presence of a hormone that regulates, in some way, the sex life.

Dr. Frederick Tilney, in his book [1] gives us an

[1] Frederick Tilney and H. A. Riley: *The Form and Functions of the Central Nervous System* (P. B. Hoeber, New York, 1921).

excellent example of a case of pineal disease. The patient, a boy of eight, had suffered from recurrent headaches since his sixth year. In his eighth year his headaches had become more severe and he had suffered from vomiting attacks. "His vision was not so good as it had been, and upon advice he began to wear glasses. During his eighth year he grew rapidly until he had reached the height of five feet three inches. He was as large as a boy of 14. In addition to his increase in stature his pubic hair made its appearance and reached full development. His external genitalia became as large as those of an adult and his sexual functions were fully established. His voice underwent transition and became much deepened. During this time, however, he suffered from repeated headaches and his vision progressively failed.

"Upon examination by an oculist he was told that he had progressive optic atrophy. He was admitted to the hospital because of his severe headaches and vomiting. At this time his vision was practically gone. Shortly after admittance to the hospital he was seized with a convulsion which lasted for half an hour. After this convulsion he never recovered consciousness but lapsed into a somnolent condition in which he remained for several weeks, at the end of which time he had a second convulsion and died a few days later.

"Upon examination at the time of his entrance

into the hospital the following observations, among others, were made: The patient gave evidence of slight loss of volitional control in both legs and arms. Although a child of eight years he looked a boy of 15 or 16 both in size and development. The mental state of the patient was difficult to estimate. He seemed precocious in certain particulars but definitely retarded in others. He had been unable to attend school because of his headaches and for this reason his actual rating could not be made.

"Furthermore, on his admittance to the hospital he was suffering from such extreme headaches that only the statement of his parents could be depended on in estimating his age. The spinal fluid on lumbar puncture appeared to be under increased tension, but it was negative to all special tests. The blood and urine were also negative. The lesion in this case was a brain tumor. Evidence of the focus of the lesion was afforded by the precocious somatic (pertaining to the framework of the body) development and precocious sexual development and growth. The optic atrophy and blindness, together with headaches, convulsions, somnolence and death, can be accounted for by a growth involving the pineal gland in such a way as to compromise the aqueduct of Sylvius (a passage which connects the third and fourth ventricle or cavity of the brain. The pineal is connected with the roof of the third ventricle) and thus give rise to an internal hydrocephalus ("water in the head"; an enlargement of the head). This explains the visual as well as the motor disturbances in the case.

"The essential clinical features of the disease are:

1. Precocious development and differentiation of the external genitalia, the premature appearance of the axillary (pertaining to the arm-pit) and pubic hair.

2. Precocious development of the sex functions.

3. Precocious abnormal growth of the long bones, producing a stature of abnormal development.

4. The appearance of signs of internal hydrocephalus, including visual disorder, headache, vomiting, with choked disk or optic atrophy.

5. The absence of all other motor or sensory symptoms."

MAMMARY GLAND

I can do no better than quote Professor Bayliss, who has critically analyzed the various views advanced as to the growth of the gland in pregnancy and the accompanying secretion of milk. "The growth of this organ is closely connected with that of the uterus (womb) in pregnancy, so that it is not surprising to find that the growth is affected by a hormone produced in the corpus luteum ("yellow body" in the ovary that grows for some time after impregnation of the ovum). . . . The second stage, associated with secretory activity in the later period of pregnancy, is independent of the corpus luteum. It has been shown by Mackenzie that the gland is not under the influence of the nervous system, but that extracts of various organs, injected into the blood current of a cat in lactation, cause secretion of milk.

"The organs found active were the pituitary body, the corpus luteum, the pineal body, the involuting uterus (the return of the uterus to normal size after child is born), and the mammary gland itself. The pituitary body is by far the most active. The fetus (the child in the womb after the third month. Before that time it is called the embryo and placenta (organ in uterus that establishes connection between mother and child) produce hormones which inhibit the gland.

"Further analysis of the action of pituitary extract was made by Hammond. The effect is said not to be due to pressing out of milk by contraction of muscle in the ducts . . . the daily yield of goats was found to be only slightly increased by injections, so that pituitary extract seems to act by setting free the constituents of milk, rather than by causing increased formation. . . ."

KIDNEY

Experiments to prove that the kidney elaborates an internal secretion have been of so conflicting a kind, that references to them would yield little information and much confusion. The curious-minded may be referred to Biedl's book (see Bibliography).[1]

---

[1] *Skin.* Lately Dr. Doege, writing in the *Wisconsin Medical Journal* (August, 1921), discusses the evidence in favor of the view that the skin has an internal as well as an external secretory function. There seems to be an intimate connection between the ductless glands and the skin. "Perhaps the most familiar examples are the appearance of myxedema with the loss of thyroid function; the dependence of certain skin eruptions or pigmentations on the sex glands, pregnancy, puberty and the climacteric period; the appearance of the bluish discoloration of the skin in Addison's disease, an affection of the adrenals. Again, the fact that many infectious diseases, such as measles, diphtheria, smallpox, spotted fever, and syphilis run their course with an essential involvement of the skin is certainly not without deeper significance, and points to the probable fact that the skin performs an important function in the overcoming of these affections. . . ."

CHAPTER X

THE RELATION OF THE DUCTLESS GLANDS
TO ONE ANOTHER

The function of the hormones generated by the ductless glands is to coördinate the various activities of the body. That there should exist a close relationship between any one ductless gland and any other or a group of others, is what might be expected, but the difficulty in proving beyond all question such relationships is great. This chapter, then, should be read with reserve; what is related here is meant to be suggestive and no more. But this chapter, like one or two others following it, will, I trust, also serve as handy summaries of much that has been discussed in previous pages.

The eminent French physiologist, Gley, writes: "The connections between the various glands are one of the fundamental facts maintained by the doctrine of internal secretions, and to deny them would be to deny a part of the doctrine of internal secretions itself. But what I criticize is the insufficiently demonstrated theory of reciprocal relations." And well he may; there are no end of

147

pitfalls that must be avoided if one is to steer clear
of hypotheses that are attractive and that are based
on the flimsiest foundation of actual knowledge.

One of the earliest attempts to give us a concrete
picture of an inter-relationship between the ductless glands was that due to the Viennese patholo-
gists, Eppinger, Falts and Rudinger who, in 1908,

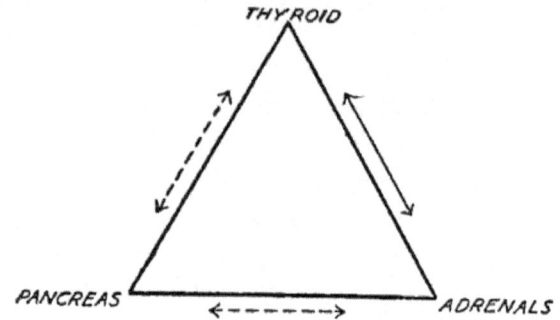

FIG. 1.   THE RELATIONSHIP OF THE PANCREAS TO THE
OTHER DUCTLESS GLANDS.

published a paper dealing with the influence of the
thyroid and the adrenals on the activity of the internal secretion of the pancreas.   It was, in fact,
an attempt to summarize our knowledge of the
modus operandi of carbohydrate metabolism.
Their views will be understood by reference to Fig.
1, and more particularly to Fig. 2.

It will be remembered that in our discussion of
carbohydrate metabolism it was stated that the
various carbohydrates, such as starch and cane

sugar, after undergoing appropriate simplification in the digestive tract are stored in the liver in the form of glycogen; and that whenever the body needs to expend energy, some of the glycogen is converted into glucose, which in turn finds its way to the muscles, where some may be resynthesized into glycogen, but where it is ultimately oxidized or

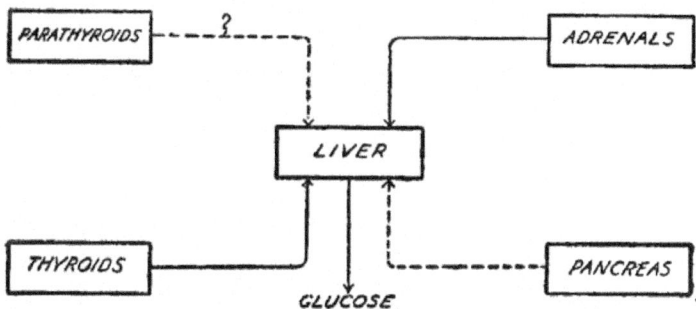

FIG. 2. THE EFFECT OF MOBILIZATION OF SUGAR IN THE LIVER.
———— = stimulation; - - - - - = inhibition
An arrow indicates the direction of action.

"burned," yielding, as final products, carbon dioxide and water. All this is a very complicated process. There must be a regulating or guiding mechanism involved to establish order in the place of chaos. We have seen how a large part of this function is taken over by the pancreas, and we have proved that it is the internal secretion of the pancreas that is responsible. We have seen, for example, how the extirpation of the pancreas removes a restraining hand from the liver, with the result that an excess of glucose appears in the blood, and finally in the urine, giving rise to the sugar disease, diabetes.

It would seem as if the thyroid, and particularly the adrenals, accelerate the conversion of glycogen to glucose, and that the pancreas, and perhaps to some extent the parathyroid, retard such a conversion. Eppinger and his colleagues reached this conclusion from studies of the effect on protein metabolism of injecting adrenaline. They found that this increased protein metabolism (the amount of protein digested and utilized) was the same as that seen in hyperthyroidism and the opposite of that produced by the removal of the thyroid gland. On the other hand, the pancreas seems to prevent the formation of an excessive quantity of sugar, for we see that such an excessive quantity is produced when the pancreas is removed. As for the parathyroid, the general feeling that it tends to neutralize thyroid activity has made some investigators class it on the side of the pancreas. These actions are shown in Fig. 2.

Fig. 1 shows that the pancreas, thyroid and adrenals, in addition to influencing sugar metabolism in the liver, influence the activity of one another. Thus the thyroid and adrenals excite one another to activity, whereas the thyroid and the pancreas, and the adrenals and pancreas, inhibit one another's activity.

These views—particularly the one referring to adrenal function—have been strenuously opposed by Gley in France and Stewart in this country. Professor Stewart's work will again be referred to in a later chapter (see p. 165). Here it may be said that he presented evidence to prove that adren-

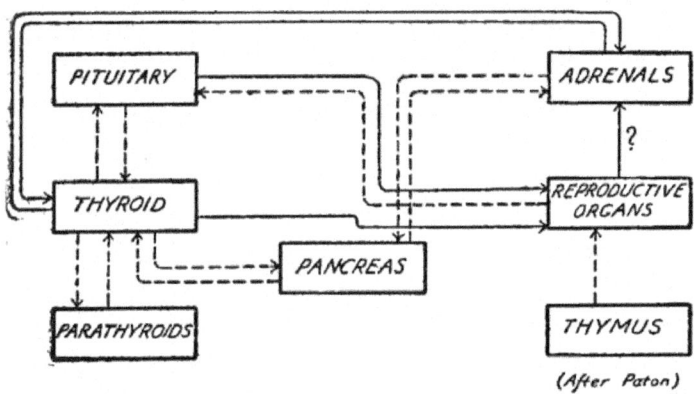

(After Paton)

FIG. 3. THE POSSIBLE INTER-RELATIONSHIPS OF THE DUCTLESS GLANDS.
———— = stimulation; - - - - - = inhibition
An arrow indicates the direction of action.

aline is not essentially concerned in experimental hyperglycemias (excess of sugar in the blood), since hyperglycemia is obtained in rabbits which have survived double adrenalectomy. It is concluded that the mobilization of sugar, of which experimental hyperglycemias are an expression, is not mediated through the secretion of the adrenals.

Fig. 3 is a more extensive diagram to show the

interrelationships of the ductless glands. Let it be emphasized again that the diagram is far from exact, because our knowledge is so incomplete. Where no connection between glands is shown, it does not follow that none exists, but that so far none has been found to exist. Where a connection is shown, we merely mean to imply that the weight of opinion is in favor of such a view, not that it is necessarily the correct view. Let us discuss the meaning of the diagram, taking up each gland in turn.

Pituitary (A) *and sexual glands.*—Hypopituitarism often gives rise to absence of secondary sexual characteristics. In a woman it may show itself by absence of pubic hair, and by arrested development of the breasts. On the other hand, castration results in an enlargement of the pituitary.

That growth depends upon the pituitary has long been known. Growth is very largely completed at puberty. This would explain why most women are smaller than men, since the former reach the stage of puberty before the latter.

In conformity with the idea that the pituitary and the sexual glands are intimately related, it has been shown that feeding tadpoles with extracts of the anterior lobe of the gland accelerates sexual development. (Goetsch.)

(B) *And thyroid.*—The removal of the thyroid

results in the enlargement of the pituitary, and vice versa. Both the pituitary and the thyroid seem to stimulate the sexual glands in the same direction. If either is removed, not only does the animal fail to grow properly, not only is there an arrested development of the mind and a tendency towards adiposity, but the sexual organs remain undeveloped.

As showing the parallelism in action of the thyroid and the pituitary in certain instances, the interesting experiment has been performed of feeding thyroidectomized rats with "tethelin," the substance isolated by Robertson from the anterior lobe of the pituitary: a beneficial effect on the growth of the animals was observed. Are we to assume that under certain conditions tethelin can take the place of the thyroid hormone?

Thyroid (A) *and pituitary.*—See under pituitary.

(B) *and sexual glands.*—Removal of the thyroid stops the growth of the sexual glands. Castration, however, does not seem to have much influence on the thyroid, though it has been stated that hyperactivity of the latter, in the shape of exophthalmic goiter, is not uncommon.

(C) *and thymus.*—Conflicting results.

(D) *and parathyroids.*—At one time the parathyroids were regarded as adjuncts of the thyroid; now the view is quite firmly established that the parathyroid is an independent organ, and that it and the thyroid may be regarded as displaying reciprocating actions. (inhibitory)

(E) *and adrenals.*—The work of Cannon and others tends to show that these two stimulate one another's activity. Hyperthyroidism, such as is found in Graves's disease, is said to increase the adrenaline in the blood, and hypothyroidism lessens the activity of the adrenals. The adrenals stimulate the production of sugar from glycogen, and it is supposed that the thyroid also acts in this way (see Fig. 2). This view advanced by Eppinger and his associates in Vienna, is based on experiments which show that the removal of the thyroid makes adrenaline less effective in bringing about glycosuria. Professor Underhill, of Yale, disagrees with this view. We shall see in a subsequent chapter (p. 165) that Professor Stewart, of Western Reserve University, has also quite a number of criticisms to offer.

(F) *and pancreas.*—We have many conflicting theories. It has been said that the diabetes produced in an animal by removing its pancreas, can be prevented by also removing its thyroid. If this unlikely view is correct, then the two glands show reciprocal actions.

Adrenals (A) *and thyroid.*—See latter.

(B) *and pancreas.*—Conflicting. Removal of the pancreas produces sugar in the urine (glycosuria). On the other hand, addition, or rather injection of adrenaline does the same. Hence the view that the two act reciprocally. Direct proof has not been forthcoming.

(C) *and sexual glands.*—Conflicting. It is said that "in cases of sexual precocity the adrenal cortex is much enlarged."

Thymus (A) *and thyroid.*—See the latter.

(B) *and parathyroids.*—Uhlenhuth, of the Rockefeller Institute, is responsible for the statement that the thymus secretes a tetany-producing substance, the action of which is neutralized by the parathyroids. This needs confirmation.

(C) *and sexual glands.*—The removal of the thymus in early life brings about the development of the sexual glands. It does seem as if the function of the thymus in early life is to retard the onset of puberty.

CHAPTER XI

THE INFLUENCE OF THE DUCTLESS GLANDS ON GROWTH AND METABOLISM

These questions have been discussed in the various portions of the book dealing with the pathology of the subject. The essential features may be brought together in this chapter.

It has been shown that castration in early life may lead to abnormal growth of the skeleton. On the other hand, the removal or atrophy of the thyroid, pituitary, and presumably the thymus, leads to arrested growth (the cretin is an example). Where the pituitary and (perhaps) the thymus are overactive, we get excessive growth (gigantism and acromegaly are types).

Whether the other ductless glands influence growth is not clear.

When we come to the influence of the glands on metabolism—on the various reactions that go on within the body—we find that the ductless glands are of great importance. The thyroid is the foremost metabolic regulator of the body. We have already seen how the metabolic rate is accelerated in hyper-thyroidism and retarded in hypo-thyroidism. It has already been shown how the pancreas, the adrenals and probably the thyroid, directly influence carbohydrate metabolism. In protein metabolism—in the assimilation and general utilization of meat, proteins of milk and eggs, for example—we have reasons to believe that the thyroid, pituitary, adrenals and the sexual glands stimulate, and the pancreas and the parathyroids inhibit such metabolism. That metabolic studies are of great value in the diagnosis of ductless glandular diseases is well illustrated in diseases connected with the thyroid (see p. 40).

CHAPTER XII

The work of Cannon of Harvard, Crile of Western Reserve University, and the war experiences of many doctors, more particularly in cases of "shell-shock," have emphasized the possibly close connection between certain phases of nervous disorder and the derangement of the ductless glands. Let us in the first instance review very briefly some of the glandular diseases that are accompanied by nervous effects, and we can then take up the work of Cannon and others in some detail.

Diseases of the thyroid come first in order. We have already seen, in discussing this subject, how a hyperthyroidism, as exemplified in exophthalmic goiter, not only increases the metabolic rate, but affects the emotions. Irritability, hasty speech, attacks of laughing and crying, and a general restlessness,—all this accompanied by a tremor, and a rapid heart beat, point to a mental as well as a physical attack. In fact, the close connection between hyperthyroidism and the mental state of the patient is such that an interesting discussion has now arisen as to whether the origin of the disease is to be laid at the door of "nerves," whose rupture plays havoc with the thyroid, or at that of the latter, which in turn affects the "nerves." An editorial in *Endocrinology* (1917) has this to say: "The work of Cannon demonstrates how complete is the cycle, and how difficult it is in a given case to ascertain whether the original cause was psychical or material. In no disease is this more evident than in Graves's disease (which, you will remember, is a common form of hyperthyroidism). Here a succession of nervous shocks may excite the adrenals until the thyroid is put into action, and hyperthyroidism arises. But again the stimulation of the vagus (nerve responsible for sensation and motion) may come from so material a source as a uterine myoma (a tumor of the womb) or other pelvic structure—as Hertzler points out. But the outcome is the same: the thyroid becomes stimulated until the threshold becomes permanently lowered."

· That the origin may be a nervous one seems reasonably clear from the experiences of the war. Dr. Cobb, a captain in the Royal Army Medical Corps, writes: "It is a well-known fact that the syndrome which we have hitherto called 'hyperthyroidism' is frequently met with among the cases of functional neuroses which arrive at the base hospitals . . . the exophthalmos is not often marked, but the fine tremor, moist skin, tachycardia (rapid action of the heart), prominent thyroid, and mental irritability are all present. The brisk reaction to any emotional excitement, with exaggeration of those features, shows that the mental element is not negligible. Furthermore, any one who has had any lengthy experience of this class of soldier-patient will agree that his mental outlook is markedly similar to that of the civilian patient with exophthalmic goiter."

In hypothyroidism, as in the cretinous child, and in the adult suffering from myxedema, we go from the extreme of rapidity in action and thought (hyperthyroidism) to complete sluggishness and mental apathy. The child coördinates poorly; it learns to talk late in years—sometimes it never passes beyond the stage of inarticulate sounds; it learns to sit and to walk late. The adult loses all reaction to strong stimuli and resembles the hibernating animal.

Professor Falta states that the English Myxedema Commission "found the apathy characteristic of myxedema to be absent in three of 109 cases. This may develop relatively early, and in the light cases may consist in a sluggishness of movement, in a retardation of the psychic functions, in an inability to form rapid conclusions, and in a slowing and monotony of speech. Magnus-Levy, the German physiologist, claims that even in the light cases the 'capability of reacting to strong stimuli' is lost. The speech may be markedly slowed, 'as if,' writes the [late] Dr. Meltzer, 'the speech mechanism were frozen in.' Charcot, the French neurologist, compares such patients to hibernating animals.

"The English Commission found among the myxedema patients investigated by them that 18 suffered from illusions, 16 from hallucinations, and 16 from frank psychosis. The psychoses belong to various types, although the melancholoid conditions predominate. The symptoms of the psychosis often develop simultaneously with those of myxedema, and vanish after thyroid therapy has been instituted, to reappear again when the therapy is discontinued."

The researches of Cannon, referred to above, tend to show that the adrenals—and more particularly one of their hormones, adrenaline,—are closely associated with the mental state of the person. The close connection between the action of adrenaline and that of the sympathetic system has been discussed in the chapter on the adrenals. We must now take up Cannon's work in some detail.

The researches of Professor Cannon.—Professor Cannon, of Harvard, working in conjunction with a number of his pupils (among whom must be mentioned de la Paz, Shohl, Wright, Washburn, Lyman, Nice, Gruber, Osgood, Gray and Mendelhall), has brought forward evidence to show that at times of emotional excitement, pain or asphyxia (suffocation), an increased secretion of adrenaline takes place. "Adrenal secretion has previously been proved to be subject to sympathetic stimulation (see page 83); and as excitement, pain and asphyxia were conditions well recognized as accompanied by sympathetic activity (manifested, for example, by inhibition of digestive functions), an attendant adrenal secretion was naturally to be expected. In a series of papers which follow the first two in 1911, experiments were described showing that adrenal secretion was serviceable in lessening muscular fatigue and in accelerating coagulation (clotting) of the blood. It was pointed out that excitement, pain and asphyxia were conditions which in natural existence would commonly be associated with struggle, and the adrenal secretion, which accompanies these three states, would be useful in great muscular effort."

Cannon criticized by Professors Stewart and Gley.—Stewart, of Western Reserve University, Cleveland (working in conjunction with his chief assistant, Dr. Rogoff), and Gley, of the Collège de France, Paris, have seriously questioned Cannon's interpretations. From his own work Stewart draws three conclusions, each one of which helps to explain a discrepancy, and all three of which tend to throw confusion into the camps of Cannon's followers: In the first place, the discharge of adrenaline is continuous; secondly, the amount of this substance in any animal is approximately constant; thirdly, the supposed variation is dependent on the rate at which the blood flows through the veins. Keeping these views in mind, Stewart finds that neither in pain, nor in asphyxia, nor in emotional excitement, can any increased secretion of adrenaline be detected.

An experiment by Professor Cannon.—Segments of rabbit intestine were placed in cylinders and these filled with samples of blood taken from the lumbo- (pertaining to the loins) adrenal veins. The blood was taken before and after stimulation of the central end of the sciatic nerve. "Normal blood removed before stimulation of the central end of the sciatic nerve caused no inhibition of the rhythmically contracting intestinal segment, whereas that removed afterwards produced a marked relaxation.[1] The conclusion was drawn

[1] In the chapter on the adrenal glands we stated that there were two accepted methods for the estimation of adrenaline.— the one a physiological, and the other a chemical method. The physiological method is the more sensitive, and for the extremely minute quantities of the substance with which we are here dealing, the physiological method is, at present, the only one that yields results. The principle employed is one which depends upon the fact that a portion of an organ, such as the uterus or the intestine, when bathed in blood or in Ringer's solution— containing a mixture of inorganic salts of a concentration similar to that found in blood—will produce rhythmic contractions that can be made to record on a slowly revolving drum; when, however, adrenaline even in the proportion of one part in one million

that the adrenal glands are affected through nervous channels when a sensory trunk (the main stem of the nerve) is strongly excited, and that they then pour their secretion into the blood stream.[1]

It must be remarked that the inhibitory influence on the beating intestinal strip is shown by adrenaline, which of course suggests that the inhibition described in this experiment is the result of an increased activity of the adrenals, with a consequent increased production of adrenaline. This adrenaline is discharged into the blood. If

is added to the solution, or when blood containing a quantity of adrenaline above the trace that is probably found normally, then the tracings on the drum show a sudden jump. For details we must refer the reader to any standard textbook of physiology. Stewart's includes a number of interesting practical exercises.

[1] For the benefit of some readers who may desire more detailed information, the following additional points in Cannon's procedure are appended: The segments of rabbit intestine were suspended lengthwise in a glass cylinder through which oxygen was passed. The segment, when not surrounded by the blood to be tested, was bathed in Ringer's solution (see above). The test blood, the cylinder and the fresh Ringer's solution were all kept at body temperature in a common bath. The blood to be tested was taken before and after the experimental procedures by passing a catheter (a tubular surgical instrument for discharging fluid from a cavity of the body) through an incision in the femoral vein (referring to the thigh) into the iliac (the haunchbone or the flank) and thence into the inferior vena cava anterior to the entrance of the lumbo-adrenal veins. A thread tied tightly around the catheter marked the point to which it was inserted and permitted reinsertion to the same point in subsequent sampling of the blood. The position of the catheter opening, which was at one side, was kept constant by attention to the position of the knot in the thread. Thus both the control blood and the blood after stimulation were taken as exactly as possible from the same region.

Professor Cannon's theory is correct there ought to be more adrenaline in the blood after stimulation—in the above experiment the stimulation was sensory—than before stimulation. His experiment tends to show that such is actually the case.

In a similar manner Cannon has shown that asphyxia "causes a change in the blood producing the same effect as adrenaline on the beating intestinal strip, namely, inhibition"; and that emotional excitement gives rise to a similar phenomenon. Hence the conclusion that stimulation of the type described—whether sensory,¹ as in pain, whether of the nature of emotional excitement, or of the nature of asphyxia—increases the secretion of adrenaline.

Stewart and Rogoff's criticism.—As we shall again refer to Professor Stewart's work, we need only mention here one or two points that bear directly on the technique employed. In a critical review of Cannon's catheter method, Stewart and Rogoff point out that the results obtained by it are only valid if the blood flow is assumed to be constant during the whole experimental period, and the method does not permit any judgment on that point. If in the course of an experiment the rate of blood flow over a particular region varies, then the samples taken at various intervals are not strictly comparable. They maintain that the secretion of adrenaline is *not* influenced by reflex stimulation, and that the only way in which the experiment would indicate an increased concentration of adrenaline in the blood is if the blood flow through the adrenal vessels were retarded.

The denervated heart as an indicator of adrenal secretion.—In this method instead of removing blood from the body, the denervated heart is used to demonstrate an increase of adrenaline in the blood. "In a cat under urethane, with vagi (nerves of sensation and motion) cut and stellate ganglia (referring to nervous matter) excised, stimulation of the central end of the cut sciatic will cause the heart rate to increase in some instances as much as 50 beats per minute. . . . The completely denervated heart can be used as an indicator of adrenal secretion in testing the influence of emotional excitement quite as well as in testing the influence of sensory stimulation and asphyxia.[1] The results obtained with the isolated heart used as an indicator of adrenal secretion thus confirm in every

---

[1] Again for the benefit of some readers certain details should be added. To denervate the heart the stellate ganglia are first removed under ether with aseptic precautions; later the right vagus nerve is severed below the recurrent laryngeal branch; and still later the left vagus nerve is cut in the neck. The heart is thus wholly disconnected from the central nervous system, and any agency causing an increase in the heart rate must exert its influence through the blood stream. With the adrenal glands normally innervated the rate was 217 per minute when the animal was calm, and 255 when excited. After the adrenal glands were removed the rate when calm was 217 and when excited 221,—an inappreciable difference.

respect the results obtained eight years ago (1911) by the catheter method." (Cannon.)

Professor Stewart objects again.—One would expect an increased rate of the denervated heart, writes Professor Stewart, when the central end of the sciatic or the peripheral of the splanchnic nerve is stimulated, for "it is obviously dependent upon the better flow through the coronary vessels"; and the increased rate of blood flow through the denervated heart increases the amount of adrenaline passing a given area in unit time. He objects to the use of any organ in the body as an indicator of adrenal secretion when asphyxia is employed as a stimulus, "because asphyxia may be expected to alter the reactivity of the test object to adrenaline, making it, for example, more sensitive." "We never supposed," he continues, "that it was possible to use in one observation an asphyxiated test object and in the comparison observation the same object with unobstructed respiration, or to assume that if there was any difference in reactions, it must be due to a difference in the rate of output of adrenaline; the conditions of the test object itself being of no moment."

When two such redoubtable adversaries as Cannon and Stewart appear in the field, it is not to be expected that either the one or the other can hope for a quick, decisive victory. Both are masters in the field of experimental physiology, and both are thinkers untrammelled by any standardized methods of thought. For every "no" of Stewart, Cannon finds a "yes." To-day as little as ever before is Cannon disposed to agree with his colleague that the adrenal effects may be accounted for on the basis of greater flow, or of altered distribution of the blood. Cannon hits back with the same weapons that Stewart employs: he does not question the results but he does question the methods. "The work of Stewart and Rogoff was admirably quantitative in character, but it was done under experimental conditions which could not afford information regarding the normal secretion of the adrenal glands or the natural conditions which affect that secretion. This conclusion applies to all inferences as to the nature of adrenal activity which they have based upon the employment of the pocket method.[1]

[1] The "pocket" refers to a pocket in the inferior vena cava. The pocket was made by opening the abdominal cavity, clamping the vena cava immediately above the iliacs, then clamping the renal veins, emptying the cava segments by stripping it upwards, and placing a clamp on the vessel above the entrance of the lumbo-adrenal veins. Any small branches of the cava segment were tied. The pocket thus formed was allowed to fill with blood from the adrenal veins, and the blood was either allowed to pass into the general circulation by removal of the clamp of the inferior vena cava, or was withdrawn and tested outside the body on preparations of rabbit uterus and intestine.

Professor Cannon's comment is characteristic: "Either because the opening of the abdomen produces a secretion unsurpassable by reflex stimulation, or because that operation abolishes abdominal reflexes, the influence of sensory stimulation on the

Professor Cannon's views as to the function of adrenaline.—This brings us to Cannon's "emergency theory." In times of stress—as when a person suffers pain, or is in an agitated state of mind —a chemical factor, in the shape of adrenaline, coöperates with nervous factors in an attempt to meet the emergency and not be overcome by it. That is why, according to Cannon, we find an increased output of adrenaline in times of stress.

But we must be careful not to misinterpret this view of Cannon's. "The concept of an emotion may be expressed either in psychological terms of subjective experience or in physiological terms of bodily change. Adrenal secretion is not essential to the subjective experience of strong emotion. Adrenaline has its effect peripherally, on outlying viscera. An assumption that subjective feeling depends on circulating adrenaline involves supporting the view that emotion as a psychological state is the consequence of visceral changes. I have, in fact, definitely argued against this view. If the critics of the emergency theory conceive emotion as bodily change, they will find in Cannon's consideration of the interrelations of emotions the point emphasized that it is the *sympathetic division* adrenal glands is not manifested. There is little wonder, therefore, that Stewart and Rogoff, who alone have employed the pocket method, with its attendant severe abdominal operation and repeated manipulation of the abdominal contents, failed to obtain the positive results which have been obtained by all other observers."

of the autonomic system which is the primary agency in mobilizing the bodily forces in times of great fear or rage. . . . These suggestions imply coördination of chemical and nervous factors, but not a dependence of the nervous factors on the chemical."

Professor Cannon's book.—Under "Bibliography" towards the end of the book I shall include sufficient references to Cannon's papers so that the reader who desires it may get inspiration from the original source; but I cannot resist the temptation of dwelling, if only for a minute or two, on his book, "Bodily Changes in Pain, Hunger, Fear and Rage," a sequel to an earlier and no less celebrated volume, "The Mechanical Factors in Digestion." It is a record of an attempt to investigate certain psychological reactions—some of which have already been discussed in these pages—by means of recognized methods in experimental physiology. The writer is strongly convinced that the great advances in psychology are not to be expected so much from the psychologist who is an arm-chair philosopher or who dabbles in "efficiency" tests, as from those men whose training has been in the experimental sciences primarily,— in chemistry, physics and biology. I do not of course mean that the philosopher has no contribution to make; I do not belong to the ultra-scientific school which takes it for granted that it, and it alone, can lay claim to the mantle of glory and achievement; but, comparatively speaking, I think that the physiologist who investigates psychological phenomena has more to offer us than the metaphysical speculator who turns psychologist.

In the early portion of the book we are introduced to the physiological methods for determining adrenaline, and are then shown that emotional excitement gives rise to adrenaline in the blood in amounts sufficient to be detected, though the amounts may be less than a few parts per million. One of the experiments dealing with emotional excitement is to bring a dog and cat near one another. The cat exposed to the barks of the dog shows an increased adrenaline output.

The fact that injection of adrenaline into the body of an animal gives rise to a glycosuria, or an increase of sugar in the urine (a form of what is commonly called "diabetes"), and the fact that emotional excitement induces an increased output of adrenaline, leads directly to the next step: can glycosuria be called forth by emotional excitement? Cannon finds that it can. One experiment is of exceptional interest. Of 25 members of the Harvard University football squad whose urine was examined immediately after a most exciting contest during the season of 1913, 12 showed sugar. After a day or two,—after a complete rest, that is— the sugar disappeared completely.

Of what value are the sugar and the adrenaline that are poured into the blood during emotional excitement? We have already indicated the answer in several portions of the book. Let Professor Cannon speak. "The adrenaline plays an essential rôle in calling forth stored carbohydrate from the liver, thus flooding the blood with sugar. . . . Since the fear emotion and the anger emotion are, in wild life, likely to be followed by activities (running or fighting) which require contraction of great muscular masses in supreme and prolonged struggle, a mobilization of sugar in the blood might be of signal service to the laboring muscles. . . . Adrenaline helps in distributing the blood to the heart, lungs, central nervous system and limbs, while taking it away from the inhibited organs of the abdomen; it quickly abolishes the effects of muscular fatigue; and it renders the blood more readily coagulable. These remarkable facts are, furthermore, associated with some of the most primitive experiences in the life of the higher organisms, experiences common to all, both man and beast—the elemental experiences of pain and fear and rage that come suddenly in critical emergencies."

In connection with these investigations, Cannon discusses and connects the excitements and energies of competitive sport; the frenzy and endurance in ceremonial and other dances; and the fierce emotions and struggles in battle.

The latter portion of the book is devoted to the nature of hunger, which has roots similar to those of fear and anger. Hunger is shown to be the direct result of contractions of the alimentary canal, a fact amply verified by the elaborate researches of .Professor Carlson, of the University of Chicago. That in fever hunger should be absent seems logical because "infection, with systemic involvement, is accompanied by a total cessation of all movements of the alimentary canal. Boldireff observed that when his dogs were fatigued the rhythmic contractions failed to appear. Being 'too tired to eat' is therefore given a rational explanation."

The closing chapter of the book pleads for moral substitutes for warfare. The key to Cannon's views is presented through the medium of William James who, in proposing a moral equivalent for war, wrote: "We must make new energies and hardihoods continue the manliness to which the military mind so faithfully clings. Martial virtues must be the enduring cement; intrepidity, contempt of softness, surrender of private interest, obedience to command, must still remain the rock upon which states are built. . . . The martial type of character can be bred without war. Strenuous honor and disinterestedness abound elsewhere.

Priests and medical men are in a fashion educated to it, and we should all feel some degree of it imperative if we were conscious of our work as an obligatory service to the state. We should be *owned*, as soldiers are by the army, and our pride would rise accordingly. We could be poor, then, without humiliation, as army officers now are. The only thing needed henceforth is to inflame the civic temper as past history has inflamed the military temper."

It is ten years since James wrote these words, and during those years much has happened to make this advice even more imperative. Neither the victorious Peace of Versailles, nor yet the Washington Conference for the Limitation of Armaments, holds out immediate relief from military dominance. One wonders what kind of calamity the gods can send us so that we may be awakened before the Great Flood sweeps us forever from off this globe.

Dr. Crile's researches.—Dr. Crile, the famous Cleveland surgeon, has advanced a theory regarding shock and exhaustion which deserves treatment in this chapter for two reasons: first, because in his theory the adrenals play an active part; and secondly because of the success he has had in the clinical application of his theory.

Shock—of which the varieties of shell-shock described during the late war are types—is characterized by a loss, to a large extent, of deliberate action. "The man in acute shock or exhaustion," writes Crile, "is able to see danger, but lacks the normal muscular power to escape from it; his temperature may be subnormal but he lacks the normal power to create heat; he understands words but lacks the normal power of response." He cannot transform *potential* into *kinetic* energy. Herein lies the key to the situation. We also see why Crile talks of the "kinetic theory of shock."

Let us dwell on this "kinetic theory" for a minute. Various stimuli arouse various associations; the latter may be of the *noci* or injurious type, or the *bene* or beneficial type. "All of life is made up of bene- and noci-associations, and the constant effort of the race and the individual is to increase the former and decrease the latter, to develop an environment which shall be as free as possible from noci-associations,—to reach a state of *anoci-association.*" *Anoci-association* is the title given to one of Dr. Crile's very suggestive books. In it he shows how the percentage of successful surgical cases may be increased by a treatment which applies the principles of anoci-association before, during and after the operation.

"The difference between normal processes and shock is that of intensity, not of kind. From these premises it becomes obvious that the exclusion of both traumatic and emotional stimuli will wholly prevent the shock of surgical operations." To accomplish the desired result, the conditions which produced the shock are ameliorated or eliminated, and the circulation is supported. For the details the reader must be referred to the book. One further quotation must suffice. "By an assuring pre-operative treatment; by the definite dulling of the nerves through the administration of a narcotic; by a non-suffocating odorless anesthetic; by a local anesthetic to cut off all afferent impulses during the course of the operation; by a second local anesthetic of lasting effect to protect the patient during the painful postoperative hours; by gentle manipulation and sharp dissection,—by the combination of all these methods—the patient is protected from damage from every factor excepting those which exist in the diseased condition from which relief is sought." [1]

[1] How successful this method is may be gauged from the fact that Dr. Crile can operate on a patient suffering from exophthalmic goiter without materially increasing his pulse rate. "It was in large measure the study of the preoperative and postoperative course of cases of Graves's disease which led to the enunciation of the kinetic theory of shock and the development of the shockless operation. . . . By the extension of employment of anociation and asepsis, the mortality in the last 6,261 operations at Lakeside Hospital (Cleveland) has been reduced to 1.6 per cent. Our series have included 58 colostomies and resections of the rectum and large intestine for cancer, with one death, and 70 resections of the stomach and gastro-enterostomies for cancer and ulcer of the stomach, with one death. Anociation and asepsis have made possible a series of 227 consecutive thyroidectomies and 180 consecutive ligations, that is, 407 consecutive thyroid operations for hyperthyroidism without a death. These are not selected cases. No patient was rejected and many were dying. Among the last 500 thyroidectomies, there have been five

Man is surrounded by noci-associations, and he is forever attempting to reach a state of anoci-association. An example of this is the attempt made by a body when infected to produce an anti-toxin. Under certain conditions—as a result of fear, worry, physical injury, infection, hemorrhage, excessive muscular exertion, starvation, insomnia— an excessive amount of energy, stored in the form of potential energy, is discharged; this "excessive conversion of potential into kinetic energy in response to adequate stimuli" leads to shock. Such is the "kinetic theory of shock," which further states that the lesions of shock are to be found in the cells of the brain, in the liver, and—what interests us most in this chapter—in the adrenals.

"In our laboratory," writes Dr. Crile, "we found cytologic changes in the adrenals in exhaustion from any cause, including insomnia; these changes being more marked in the cortex than in the medulla. Apparently adrenaline alone can cause the brain greatly to increase its work. By cross-circulation experiments, we have found that adrenaline causes increased activity of the central vasomotor mechanism. Not only can adrenaline, as Cannon has shown, cause all the basic phenomena of exertion, emotion, infection, etc., but it also causes brain cell lesions identical with those produced by

deaths, a mortality rate of one per cent., and among the last 500 ligations, two deaths."

exertion, emotion, infection, etc. . . . The injection of adrenaline causes an immediate increase in the conductivity of the brain to above normal, followed by a later decrease to below the normal; moreover, adrenaline causes an immediate increase in the temperature of the brain, as evidenced by thermo-couple measurements."

The work of Drs. Stewart and Rogoff.—In discussing Dr. Cannon's work we have also referred, a number of times, to that of Drs. Stewart and Rogoff, if only because the latter two are equally eminent authorities, and mainly because they combat much of what Cannon has to say. The nature of this work and the limitations of space will prevent us from giving details, but the reader who is interested may refer to the list of references to their work that is included in the bibliography.

Adrenaline is the big theme. Is there or is there not an increase of this substance when an animal is emotionally aroused? We have already suggested that Stewart's answer is in the negative, and that he explains Cannon's positive results by finding fault with the method employed, and with the way in which the results were interpreted.

It will also be remembered that Cannon makes much of the fact that not only does adrenaline give rise to a hyperglycemia, but—what might be expected if what Cannon asserts is true—so do the emotions when aroused sufficiently; for, according to Cannon, the emotions arouse the adrenals to particular activity, an increased quantity of adrenaline then appears in the blood, and this in turn interferes with normal carbohydrate metabolism to such an extent that an abnormal quantity of sugar finds its way into the blood, giving rise to hyperglycemia.

"We have recently," write Drs. Stewart and Rogoff, "studied the question whether adrenaline secretion of the adrenals is indispensable for the production of certain experimental hyperglycemias. The majority of previous investigations have suffered from the defect that they were carried out, if not on practically moribund animals, at least on animals still under the effects of a serious operation. This undoubtedly is the chief reason for the astonishing lack of uniformity in the results. Working with animals (cats) in which the adrenaline secretion was abolished or reduced to an insignificant fraction of the normal by removal of one adrenal and section of the nerves of the other (an operation which does not preclude the continued life of the animal in good health), we were able to show that two forms of experimental hyperglycemia—that produced by ether and that produced by asphyxia—are as readily obtained in the absence of adrenaline secretion as when the adrenals have not been interfered with." What, then, have the adrenals got to do with hyperglycemia?

Neither have Drs. Stewart and Rogoff less sharp criticism to offer with regard to the alleged connection of adrenaline with the condition known as "shock," a subject we have discussed in the last few pages. "A large though quite undeserved place has been occupied in clinical literature of shock and allied conditions by adrenal insufficiency, or one or other of its aliases. There is no evidence that any notable change occurs in the adrenaline output in either direction." The experimental methods of producing shock in dogs and cats—by exposing and manipulating the intestines, by partial occlusion of the inferior vena cava, by hemorrhage and by "peptone" injection—led to a permanent lowering of blood pressure; but the rate of output of adrenaline after the blood pressure had been permanently lowered was found to be the same as before the lowering of the blood pressure, "within the limits of error of the methods used for assaying the adrenaline."

Professor Stewart's caustic pen hits at the "clinical endocrinologist" even more than at the experimental physiologist. He writes: "In reading the papers by 'clinical endocrinologists,' especially the French and Italians, the physiologist can scarcely escape the feeling that here he has broken through into an uncanny fourth dimension of medicine, where the familiar canons and methods of scientific criticism are become foolishness, where fact and hypothesis are habitually confounded, and 'nothing is but what is not.' "

Such criticisms, coming from a critic whose aim is to create rather than to destroy, are of the utmost value to the progress of the science. It deserves prominent place in a book such as this, where the attempt is made never to confound fact with fancy.[1]

The sexual glands and the nervous system.—That the sexual glands and the nervous system are closely related is only too obvious from the many studies on the sex problem, and by the more direct method of castration in animals and men. Much of all this has already been discussed in another chapter (page 94). Less pronounced connections between the glands and the nervous mechanism are noticeable in diseases of the pituitary and the parathyroids.

[1] I have just (December, 1921) returned from the annual meeting of the Federation of American Societies for Experimental Biology, held this year at Yale University. These meetings were much enlivened by papers by Stewart and Cannon. Both are not only first-class scholars but excellent debaters, and their annual tilts are eagerly looked forward to. This eagerness on the part of the onlookers to enjoy the fun brought forward this remark from Dr. Stewart: "We are not waiting for Dr. Cannon to say something and then to jump at him; we merely seek the truth, just as I know he does." This but brought laughter and knowing looks. Dr. Carlson, the Chicago physiologist, whose sympathies are evidently more with Stewart than with Cannon, brought down the house with this remark: "I am glad to find that Cannon no longer pins his faith to the adrenals alone; for that he and the Society are to be congratulated on a return to 'normalcy.' "

Psycho-analysis.—It would be strange that in the treatment of those most complex of phenomena that group themselves under mental disorders, conflicts should not arise between the enthusiastic endocrinologist and the Freudian disciple; the one with his eye to the ductless glands as the source of much evil, and the other with his emphasis on the repression of the emotions. Professor Cushing, in an article on "Psychic disturbances associated with disorders of the ductless glands," leans towards the former school, though he is not blind to the merits of psycho-analysis. He writes: "The various neuroses (nerve diseases) and asthenias (loss of strength) may result primarily as the result of some disturbance of internal secretion which paves the way for the dreams, symbolisms and other manifestations dissected by the psycho-analyst. . . . It is quite probable that the psycho-pathology of every-day life hinges largely upon the effect of ductless gland discharges upon the nervous system. This is particularly worthy of consideration in the study of child psychology in its relation to puberty and adolescence, especially in those individuals in which there is some underlying, possibly inherited, functional deviation in the chemistry of the internal secretion. . . ."

Shell-shock.—This brings us to the last phase of our subject, that of shell-shock. (The reader is advised to re-read Dr. Crile's views on shock, page

174.) We may perhaps preface our remarks by saying that the strenuous life led by the individual in the city, especially by the "hustler," and especially by the "hustler" with a weak resistive capacity, may cause a nervous breakdown which is not far removed from shell-shock. Again we must refer the reader to Dr. Crile.

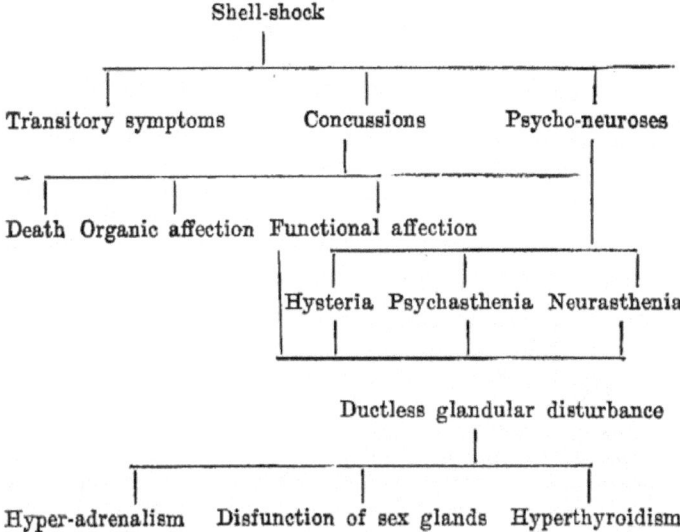

Shell-shock has been defined as "the condition which follows exposure to the forces generated by the explosion of powerful shells in the absence of any visible injury to the head or spine." It may result in a speedy recovery, in a definite concussion, or in the appearance of symptoms of the psycho-neuroses. If either of the latter two symptoms makes its appearance, it may be in connection with a disturbance of one or more of the ductless glands.[1] Dr. Cobb has drawn up an attractive classification which is given above.

The classification of nervous disorders is due to Beard, Charcot, Möbius and Janet. Janet divided psycho-neuroses as shown in the table. "Hysteria" represents a typical mental disintegration, whereby there is a splitting of mental processes, so that two separate and unconnected streams exist in the mind (for example, anesthenia, amnesia [loss of memory], paralysis); "psychasthenia" refers to cases characterized by phobias (persistent insane dread or fear), hesitations, doubts, anxieties; while "neurasthenia" refers to cases showing a preponderance of symptoms referable to physical exhaustion (fatigue, indigestion, disturbances of excretion, etc.).

[1] Clemence Dane, the author of "A Bill of Divorcement," now playing in New York, has drawn a powerful and moving picture of the shell-shocked soldier. Allan Pollock, who interprets the part, was himself through this living hell. The interpretation is a triumph of the actor's skill.

CHAPTER XIII

The treatment of ductless glandular diseases by means of extracts of appropriate glands dates back to Brown-Séquard's investigations in 1889, and Brown-Séquard merely revived the old "humoral" view of disease. Hippocrates advocated the efficacy of various organs, and Hahnemann, the originator of homeopathy, built up a subdivision of the subject, isopathy, which dealt with diseased organs, and with their cure by the administration of fresh organs. Is your liver out of joint? Then we will prescribe the liver of a wolf. Have you a tremor? The brain of a hare will put you on your feet again. Are you a sufferer from dyspepsia? Take the lung of a fox and you will get well. Is your stomach misbehaving? Take rennin.

The "humoral" philosophy throve, made many converts, did some good and very much mischief, and gradually died out. Brown-Séquard brought it to light again; but, let us add, in quite a modern form, and with reasons for its revival drawn from the knowledge of the nineteenth and not the ninth

185

century. One may truly say of him that he is the founder of the conception of ductless glandular function as we understand it to-day.

Poor Brown-Séquard! In glorious company with other ill-received or unrecognized geniuses, he became the laughing-stock of scientific Paris. He described to his audience how he had administered to himself testicular extracts, and how, as a result of this administration, his vigor and youthful desires and appetites had returned. (In 1889 when this announcement was made Brown-Séquard was 70 years old.) The Academy laughed, and Paris and the other capitals of Europe made the most of a sensational piece of news.

Was there any foundation for Brown-Séquard's claim? From what we know to-day, not very much. Not even the most enthusiastic exponents of "rejuvenation" by means of the sexual glands advocate such a procedure.

If the use of testicular extract proved discouraging, it did not prevent the use of extracts from other glands that produce an internal secretion. And then came the truly remarkable discovery that in myxedema and cretinism, examples of hypothyroidism, the administration of thyroid extracts brought cures—cures that lasted, to be sure, only so long as treatment with the extract was continued. This discovery made those who had scoffed at Brown-Séquard revise their opinion of that illustrious Frenchman. A new impetus was given the subject, and glandular treatment became the hope of a world full of maladies.

Despite an enormous amount of work, it cannot be said that we have accomplished much with glandular extracts beyond their use in thyroid disease. Pituitary extracts have had a measure of success in pituitary disorders, but not comparable to the corresponding extract from the thyroid when applied to cases of hypo-thyroidism.

With a persistence worthy of some admiration, glandular advocates, failing to get results with any one extract, tried each one of the others in turn. Still without result, they adopted pluriglandular treatment—that is to say, treatment by the use of extracts from several glands. It cannot be said that these methods yielded any better results.

Now the question may very naturally be asked, why should thyroid extract be of service in hypothyroidism, and why should extracts from other glands be of so little service in other diseases due to glandular insufficiency? We do not know, though we can speculate as to the cause. Are we to assume that only the thyroid hormone is resistive enough to escape all dangers and reach its destination safely, whereas other hormones are destroyed on the way? [1] This is not a likely hypothe-

[1] "No assumption is needed in this case any longer, because we now know that thyroxin is of a very different nature from any

sis, for even injecting directly into the blood—
the path along which the hormones travel—does not
materially improve matters. Or perhaps extracts
other than the thyroid contain substances in addition to the hormones that are in themselves injurious to the system. Unfortunately for this theory,
we know that adrenaline, 100 per cent. pure, representing the adrenal hormone, does not cure Addison's disease, a disease of the adrenals. Of course
it may be said with much force that adrenaline is
the hormone of but one portion of the adrenals, the
medulla, and that the hormone of the cortex of the
gland has not yet been isolated; so that until this

other of the constituents of the ductless glands. We know that
it does not vary in its amount in the tissues except in a minor
degree, that it is constantly supplied and that any single portion
of it functions for as long as seven weeks after administration
or after the gland manufactures it; that is, it acts as a catalyst,
and it does not come under your definition of a hormone. It
acts in a manner to increase the rate of oxidation within the
tissues, and we can now picture the chemical changes occurring
which permit thyroxin to function as a catalytic agent. None of
the active constituents of the other ductless glands acts in this
way in respect to time. Adrenaline does undoubtedly act as a
catalyst, increasing the rate of oxidation within the cells, but it
functions for a very short period of time. That, to my mind,
is the explanation of why the adrenal and the hypophysis are so
closely related to the nervous system. They are strictly emergency glands and their output must be increased and decreased
on demand. They function for a brief interval and that is why
it is impossible to administer them successfully, because only
relatively massive doses are given, interspersed with periods of
zero administration. Thyroxin is just the reverse of this. It
does not function immediately and lasts for as long as five to
seven weeks, so that it is not only unique among the glands for
therapeutic purposes, but it is unique in its chemical properties." (E. C. Kendall.)

is done, and until the *pure* hormones from both parts of the gland are used, it would be premature to draw any conclusions.

Perhaps the most plausible suggestion so far advanced is that in the preparation of a glandular extract, involving physical and chemical processes, the chemical configuration of the hormone is possibly altered, and hence its physiological action becomes lost. To point to the thyroid extract as disproving this hypothesis does not hold, for the answer may be made that there is no reason to suppose that all the hormones have the same degree of resistive power. The antiscorbutic vitamine is more easily destroyed than the antirachitic; why may this not be true of hormones as well? Why may not one hormone fall a prey to chemical agents more quickly than another?

Unquestionably the next step in our forward march will be the isolation, in a pure form, of the hormones from the pituitary, the cortex of the adrenals, the sexual glands, etc. Until this is done we can hope little more from glandular treatment than what has already been accomplished.[1]

[1] At the recent (Dec., 1921) meeting of the Physiological Society, held in New Haven, Professor Macleod, of the University of Toronto, read a paper on the value of pancreatic extracts, in which he pointed out that the blood sugar of a depancreatized dog could be lowered by injecting a pancreatic extract, and that neither an extract of the liver nor one of the spleen had that effect. Though preliminary in character, the investigation is important and suggestive.

CHAPTER XIV

PLANT HORMONES

The work of Professor Bottomley, of the University of London, has made it very probable that vitamines play an important part in the plant, as well as in the animal kingdom. He has even presented some very good evidence to show that the vitamines found in the animal world can be traced to vegetable sources; that though the animal needs vitamine, yet it cannot synthesize it, but must rely on this synthesis being accomplished by plant cells. This work of Bottomley's has been amply confirmed in many quarters.

Similar problems present themselves when dealing with hormones. In the first place, are there plant hormones that are activators in the sense that the animal hormones are? And is it possible that the hormones in our body are synthesized in the plant, rather than in the animal kingdom?

The second question will be disposed of first. We know that glands are factory centers of the body. We know that the gland has the power of taking various materials from the blood, and converting them into an entirely new product, or products. The essential product of a ductless gland is its hormone—at least, that is the supposition—and this is probably manufactured by the gland. The indirect proof we have for this statement is that under normal conditions the hormones in the body are sufficient in amount, or sufficiently active, to perform their specific functions; and it is only when a pathological condition sets in that the hormonic function is disturbed.

You may say that this is not very convincing. You may claim that under normal conditions the gland has the power of removing the hormone from the blood, and that it no longer has such power under pathological conditions. This objection can, I think, be met with in this way: If we examine the constitutional formula for adrenaline, a hormone in the adrenals, or of thyroxin, a hormone in the thyroid gland, we shall see that no food we eat contains any such substance. We do, however, find that their formulas show them to be closely related to certain substances that are included in our diet—or at least, are formed in the digestive tract as a result of the food we eat. For example, the amino-acid tryptophane [1] shows certain resemblances to thyroxin; hence the opinion that one of the possible reasons why tryptophane is

[1] See the chapter on Amino-Acids in the author's book on Vitamines.

an essential acid is that the body needs it for the manufacture of thyroxin; though so far as I know no one has as yet shown that an increase of tryptophane in our diet increases the quantity of thyroxin in the body.

Now as to the next question: are there plant hormones? And if so, do they play a part in the plant kingdom analogous to that played by hormones in the animal kingdom? Let us quote Professor Bayliss on this point:

"Although there is no such effective way of chemical interchange in plants as there is in the circulating blood of animals, there is distinct evidence that chemical products of one part are able to influence the activities of other parts. The lateral roots, which normally grow horizontally, can be made to grow vertically downwards if the main root is removed. Errera investigated, in pines, the corresponding change of direction of growth of a branch into a vertical stem when the apical bud of the main stem is removed. He suggested that the apical bud of the main stem forms some kind of internal secretion, which prevents the upward growth of the lateral shoots as long as this apical bud is present.

"Keeble considers that such 'chemical stimulators' play a part in the transfer of activity of localized cambium cells to others in their neighborhood. In the case of *Convoluta Roscoffensis,* the signal for the commencement of the later phases of development owes its origin to the presence of the green algal cells, without whose concurrence, probably by the production of a hormone, no kind of artificial feeding has been found effective.

"Mention may also be made of the substance extracted by rain from grass, which has been shown by Pickering to be injurious to apple trees. They should not, in fact, be surrounded by growing grass, as is common in orchards."

We know that a number of ethereal salts, or, as the chemist calls them, "esters," act as accelerators, in the sense of hastening the flowering of plants and the ripening of fruits; and we know further that the essential oils present in plants are largely made up of such "esters." As to whether these essential oils are manufactured by glands similar to those existing in the animal kingdom, and as to whether there is any interrelationship between such glands, cannot, at present, be answered. It is supposed that these "esters," and substances other than "esters" that act like accelerators, stimulate enzyme (ferment) action in the plant, particularly in the later stages of its development. On this basis the many color changes that take place during the ripening period, the autumnal color of leaves, and the dropping of the leaves from the stem of the

plant, have been explained; but all this is, at present, largely fancy and little more.[1]

[1] Mention may here be made of a more direct experiment by
Dr. Budington, of Oberlin College. He experimented with sound
onion bulbs which were placed in a nutrient solution to which
small quantities of glandular extracts—from the thyroid, the
adrenal and the pituitary—were added. A certain amount of
"retardation" in the growth of root tips when thyroid extract
was used, and marked modifications of growth when iodine in
the form of potassium iodide was used, were obtained. "While
no general conclusion can be based on experiments limited to a
single form, the indication is that thyroid constituents—and it
will be remembered that iodine is such a constituent—may influence the rôle of protoplasmic action in cells other than those
of animal tissues."

REFERENCES

REFERENCES

GENERAL. There are a number of books dealing with the glands of internal secretion. A standard work is that by A. Biedl: *The Internal Secretory Organs* (William Wood & Co., New York). More than 100 of the 600 pages are devoted to references to the original literature. Another excellent work is S. Vincent's *Internal Secretion and the Ductless Glands* (Edward Arnold, London). C. E. de M. Sajous' *The Internal Secretions and the Principles of Medicine* (F. A. Davies Co., Philadelphia) is an ambitious work in two volumes. E. A. Schafer, the Edinburgh physiologist, is the author of *The Endocrine Organs* (Longmans, Green & Co., London), which emphasizes the physiological rather than the clinical point of view. On the other hand, W. Falta's *The Ductless Glandular Diseases* (P. Blakiston's Son & Co., Philadelphia) is wholly clinical. D. Noël Paton in *The Nervous and Chemical Regulators of the Body* (Macmillan & Co., London) emphasizes chemical factors. For those having a reading knowledge of German, A. Weil's *Die Innere Sekretion* (Julius Springer, Berlin) may be recommended, since it is both authoritative and up-to-date. A very good historical development is given by E. Gley in *The Internal Secretions* (Paul B. Hoeber, New York; translated from the French by Maurice Fishberg). A more recent book by the same author, who is professor of physiology at the

Collège de France, may be suggested to those possessing a working knowledge of the French language; it is entitled *Quatre Leçons des Sécrétions Internes* (J. B. Ballière et Fils, Paris). Other books dealing with internal secretions are I. G. Cobb: *The Organs of Internal Secretion* (William Wood & Co., New York); S. W. Bandler: *The Endocrines* (W. B. Saunders Co., Philadelphia); L. Berman: *The Glands Regulating Personality* (Macmillan & Co.); and H. R. Harrower: *The Internal Secretions in Practical Medicine* (Chicago Medical Book Co., Chicago).

Among medical encyclopedias that include articles on internal secretions, two may be mentioned because they are recent productions. An article on ductless glands will be found in volume 3 of *The Oxford Medicine* (Oxford University Press, London), and another, in volume 3 of *The Nelson Loose-Leaf Living Medicine* (Nelson & Co., New York).

Books dealing with various phases of medicine include chapters on the ductless glands. Some of these are J. J. R. Macleod: *Physiology and Biochemistry in Modern Medicine* (C. V. Mosby, St. Louis); R. Burton-Opitz: *A Text-Book of Physiology* (W. B. Saunders & Co., Philadelphia); A. P. Mathews: *Physiological Chemistry* (William Wood & Co., New York); L. Luciani: *Human Physiology*, volume 2 (Macmillan & Co., London); W. M. Bayliss: *The Principles of General Physiology* (Longmans, Green & Co., London); W. H. Howell: *A Text-Book of Physiology* (W. B. Saunders & Co., Philadelphia); W. G. MacCallum: *A Text-Book of Pathology* (W. B. Saunders & Co., Philadelphia); W. Osler: *The Principles and Practice of Medicine* (D. Appleton & Co., New York); E. H. Starling:

Principles of Human Physiology (Lea & Febiger, Phila-
delphia); G. N. Stewart: *A Manual of Physiology* (Wil-
liam Wood & Co., New York); M. Kahn: *Functional
Diagnosis* (W. F. Prior Co., Hagerstown, Maryland); and
H. G. Wells: *Chemical Pathology* (W. B. Saunders & Co.,
Philadelphia).

VITAMINES AND HORMONES. The possible relationship
or identity of these two substances has been urged, more
particularly, by Voegtlen and Myers, and by Dutcher
(*American Journal of Physiology,* volume 49, page 124,
1919; *Journal of Pharmacology and Experimental Therapeutics,* volume 13, page 301, 1919; and *Journal of Biological Chemistry,* volume 39, page 63, 1919). This theory has met with opposition; see, for example, Anrep and
Drummond's paper on "The Supposed Identity of the
Water-Soluble Vitamine B and Secretin" (*Journal of
Physiology,* volume 54, page 249, 1921). A possible connection between vitamine B and adrenaline has been suggested by MacCarrison (see the *Indian Journal of Medical
Research,* volume 6, pages 275 and 550, 1919, and the
Proceedings of the Royal Society, section B, volume 91,
page 103, 1920). Kellaway's paper on "The Effect of
Certain Dietary Deficiencies on the Suprarenal Glands"
(*Proceedings of the Royal Society,* section B, volume 92,
page 6, 1921) should also be consulted.

THE THYROID. (See also the "general" references
above.) A number of books dealing more specifically with
this gland have been published. R. McCarrison's *The
Thyroid in Health and Disease* (Ballière, Tindall & Cox,

London) may be especially recommended. A. Crotti is the author of the *Thyroid and the Thymus* (Lea and Febiger, Philadelphia). Other books are H. J. Ochsner: *Surgery and Pathology of the Thyroid and Parathyroid Glands* (C. V. Mosby, St. Louis) ; and H. Richardson: *The Thyroid and Parathyroid Glands* (P. Blakiston's Son & Co., Philadelphia).

Kendall's work on the "Isolation of the Iodine Compound Which Occurs in the Thyroid" will be found in the *Journal of Biological Chemistry,* volume 39, page 125, and volume 40, page 265, 1919. His lecture before the Harvey Society ("The Chemistry of the Thyroid Secretion," the *Harvey Lectures,* 1919-1920, published by J. B. Lippincott Co., Philadelphia) was a fine historical review of the entire work.

The subject of hyper-thyroidism has received considerable attention lately. In this connection the reader will find in W. M. Boothby's article, "Adenoma of the Thyroid with Hyper-thyroidism" (*Endocrinology,* volume 5, page 1, 1921), a very thorough discussion of the types of thyroid disease, with references to the latest literature. H. R. Harrower's booklet, *Hyperthyroidism* (Glendale, California), suffers from an over-enthusiastic treatment. The favorable effects of surgical treatment coupled with "physiologic rest" are advocated by C. W. Crile in "Surgery Versus Roentgen Ray in the Treatment of Hyperthyroidism" (*Journal of the American Medical Association,* volume 77, page 1324, 1921).

With regard to the subject of metamorphosis, a very exhaustive review, with references to the original literature, is given by J. F. Fulton in his article, "The Controlling

Factors in Amphibian Metamorphosis" (*Endocrinology,* volume 5, page 67, 1921). A shorter review is that by L. T. Hogben (*Science Progress,* volume 15, page 303, 1920). Jacques Loeb's "Natural Death and the Duration of Life," an article included in B. Harrow's *Contemporary Science* (Boni & Liveright, New York), suggests many fascinating possibilities.

How the determination of the basal metabolic rate helps to diagnose the various types of thyroid disease is discussed by a number of physicians in a forthcoming volume, *Basal Metabolism* (Sanborn Co., Boston). There are any number of individual articles dealing with basal metabolism; only a few of these can be mentioned here. See, for example, F. G. Benedict, *Journal of the American Medical Association,* volume 77, page 247, 1921; J. H. Means, *Journal of the American Medical Association,* volume 77, page 347, 1921; W. M. Boothby, *Journal of the American Medical Association,* volume 77, page 252, 1921; F. H. Lahey, *Boston Medical and Surgical Journal,* volume 184, page 348, 1921; J. R. Murlin, *Science,* volume 54, page 196, 1921; C. W. McCarthy, *Journal of the American Medical Association,* volume 76, page 978, 1921. In volume 1 of the *Oxford Loose-Leaf Medicine* (Oxford University Press, London) will be found Du Bois' article on "Clinical Calorimetry Methods of Study of Metabolism."

With regard to the prevention of simple goiter in man, Marine, and his co-worker Kimball, are widely known for such studies. Some of their papers are to be found in the *Journal of the American Medical Association,* volume 77, page 1068, 1921, and volume 73, page 1874, 1919; *Ohio State Journal,* October, 1920; *Archives of Internal Medi-*

cine, volume 22, page 41, 1918, and volume 25, page 661, 1920; and *Journal of Laboratory and Clinical Medicine,* volume 3, page 40, 1917. A suggestive article by E. R. Hayhurst, entitled "The Present-Day Sources of Common Salt In Relation to Health and Especially to Iodine Scarcity and Goiter" (*Journal of the American Medical Association,* volume 78, page 18, 1922), wherein it is urged that "common salt for dietary purposes should include not only sodium chloride but also sodium iodide," should be consulted.

THE PARATHYROIDS. (See also the "general" references above.) Two books that deal with the parathyroids as well as with the thyroid have already been mentioned: H. J. Ochsner: *Surgery and Pathology of the Thyroid and Parathyroid Glands* (C. V. Mosby, St. Louis), and H. Richardson: *The Thyroid and Parathyroid Glands* (P. Blakiston's Son & Co., Philadelphia). An exhaustive account of these glands, accompanied by a very complete bibliography, may be found in W. M. Boothby's "The Parathyroid Glands" (*Endocrinology,* volume 5, page 403, 1921). MacCallum and Voegtlin (*Journal of Experimental Medicine,* volume 11, page 118, 1909) take up the question of the relation of tetany to the parathyroid glands and to calcium metabolism.

THE PITUITARY GLAND. (See also the "general" references above.) The classical work in English is Harvey Cushing's *The Pituitary Body and Its Disorders* (J. B. Lippincott & Co., Philadelphia). W. Blair Bell is also the author of a book on the pituitary (William Wood & Co.,

New York). T. B. Robertson's work on the isolation of a substance from the anterior lobe of the pituitary is set forth in an article in *Endocrinology* (volume 1, page 24, 1917).

THE ADRENAL GLANDS. (See the "general" references above, as well as the references listed under "the nervous system and the ductless glands.") Fine critical articles are those by Stewart and by Barker (*Endocrinology*, volume 5, page 283, 1921, and volume 3, page 253, 1919). An account of the chemistry of adrenaline is given by Abel (*Johns Hopkins Hospital Bulletin*, volume 9, page 215, 1898, and volume 12, page 80, 1901), Takamine (*American Journal of Pharmacy*, volume 73, page 523, 1901; *Journal of Physiology*, volume 27, page xxix, 1901) and Friedmann (*Beiträge zur chemische Physiologie*, volume 8, page 95, 1906). Barger's volume, *The Simple Natural Bases* (Longmans, Green & Co., London) should also be consulted.

THE ORGANS OF REPRODUCTION. (See also the "general" references above.) A standard work is Marshall's *Physiology of Sex Reproduction* (Longmans, Green & Co., London). Steinach's book, *Verjüngung* (Julius Springer, Berlin), gives an excellent summary of the author's researches. Phases of Steinach's work are discussed in the *Journal of the American Medical Association* for Jan. 29, 1921 (page 348), Aug. 14, 1920 (page 490), Aug. 28, 1920 (page 617), Sept. 11, 1920 (page 755), and Dec. 25, 1920 (page 1811). Readers with a knowledge of German may be referred to Stieve's

*Entwicklung, Bau and Bedeutung der Keimdruesen-
zwischenzellen* (development, structure and significance of
the interstitial cells of the gonads) (J. F. Bergman,
Munich). Voronoff gives an account of his experiments in
the book entitled *Life* (E. P. Dutton & Co., New York).
An analysis of the behavior of organs after transplantation
is discussed by L. Loeb (*Journal of Medical Research,*
volume 39, page 189, 1918). An editorial in the *Journal
of the American Medical Association* (volume 75, page
1070, 1920) should also be consulted. J. S. Horsley is
the author of an instructive article on the "Suturing of
Blood Vessels" (*Journal of the American Medical Association,* volume 77, page 117, 1921).

THE PANCREAS AND THE LIVER. (See also the "general"
references above.) There is a mass of literature on these
organs; we can refer to but one or two recent articles.
See, for example, E. P. Joslin: *Diabetes Mellitus* (Lea &
Febiger, Philadelphia); F. M. Allen: *Diabetes Mellitus*
(Physiatric Institute, Morristown, N. J.); E. P. Joslin:
"The Prevention of Diabetes Mellitus" (*Journal of the
American Medical Association,* volume 76, page 8, 1921),
wherein we find the cry that diabetes is a penalty for
obesity; J. J. R. Macleod: "The Sugars of the Blood"
(*Physiological Reviews,* volume 1, page 208, 1921), a
comprehensive review of the significance of blood sugar;
and E. Langfeld's series of articles on the significance of
a physico-chemical factor, the hydrogen ion content, in the
regulation of blood sugar (*Journal of Biological Chemistry,* volume 46, pages 381, 393, 403, 1921).

THE INTESTINAL HORMONE. (See also the "general" references above.) Two of the papers by Bayliss and Starling are "The Mechanism of Pancreatic Secretion" (*Journal of Physiology*, volume 28, page 325, 1902) and "The Chemical Regulation of the Secretory Process" (*Proceedings of the Royal Society*, section B, volume 73, page 310, 1904). The books on physiology by Bayliss and by Starling, referred to under "general" references, give good accounts of the discovery and action of secretin.

THE THYMUS, SPLEEN, MAMMARY GLAND, PINEAL AND KIDNEY. (See the "general" references above.) Quite a number of investigators are busying themselves with the problem of the function of the thymus. See, for example, J. A. Hammar (*Endocrinology*, volume 5, page 543, 1921), E. Uhlenhuth (*Endocrinology*, volume 3, page 284, 1919), and M. B. Gordon (*Endocrinology*, volume 2, page 405, 1919). "Is there a thymic hormone?" asks Hoskins (*Endocrinology*, volume 2, page 241, 1918); he is inclined to answer in the negative. See, also, Crotti's book, *The Thyroid and the Thymus* (Lea and Febiger, Philadelphia), and an editorial in the *Journal of the American Medical Association* (volume 77, page 2063, 1921).

N. B. Eddy (*Endocrinology*, volume 5, page 461, 1921) reviews the functions of the spleen. Sir Berkeley Moyniham (W. B. Saunders Co., Philadelphia), and Pearce, Krumbhaar and Frazier (J. B. Lippincott Co., Philadelphia) are authors of books dealing with the spleen.

Two articles on the pineal, that also include the literature, are Horrax's "Studies on the Pineal Gland" (*Ar-

chives of Internal Medicine, volume 17, page 607, 1916) and Bailly and Jeliffe's "Tumors of the Pineal Gland" (*Archives of Internal Medicine,* volume 8, page 851, 1911). See, also, W. E. Dandy's "The Treatment of Brain Tumors" (*Journal of the American Medical Association,* volume 77, page 1853, 1921). Tilney and Riley's *The Form and Functions of the Central Nervous System* (P. B. Hoeber, New York) contains a mass of valuable material.

THE RELATION OF THE DUCTLESS GLANDS TO ONE ANOTHER. (See also the "general" references above.) Certain angles of this problem are taken up by Stewart and Rogoff (*American Journal of Physiology,* volume 46, page 90, 1918), and Anon. (*Endocrinology,* volume 1, page 404, 1917).

THE INFLUENCE OF THE DUCTLESS GLANDS ON GROWTH AND METABOLISM. (See the "general" references above.) Biedl discusses this question in an article entitled "The Significance of the Internal Secretions in Disturbances of Metabolism and Digestion" (*Endocrinology,* volume 5, page 523, 1921).

THE NERVOUS SYSTEM AND THE DUCTLESS GLANDS. (See also the "general" references above.) A mass of literature, much of it of a pseudo-scientific nature, has accumulated on this phase of the subject. Only a few of the books and pamphlets will be referred to. See, for example, M. Laignel-Lavastine: *The Internal Secretions and the Nervous System* (Nervous and Mental Disease Publishing Co., New York); W. Langdon Brown: *The Sympathetic Nervous System in Disease* (Oxford University Press, London); W. B. Cannon: *Bodily Changes in Pain, Hunger, Fear and Rage* (D. Appleton & Co., New York); G. W. Crile: *A Physical Interpretation of Shock, Exhaustion and Restoration* (W. B. Saunders Co., Philadelphia); G. W. Crile: *Anoci-Association* (W. B. Saunders Co., Philadelphia); F. W. Mott: *War Neuroses* (Oxford University Press, London); W. Harris: *Nerve Injuries and Shock* (Oxford University Press, London); T. R. Elliot: "Ductless Glands and the Nervous System" (*Brain*, volume 35, page 306, 1913); N. Pende: "Endocrinopathic Contributions to Pathology" (*Endocrinology*, volume 3, page 329, 1919); Y. Henderson, H. W. Haggard and R. C. Coburn: "The Acapnia Theory, Now" (*Journal of the American Medical Association*, volume 77, page 424, 1921); C. W. Crile: "The Mechanism of Shock and Exhaustion" (*Journal of the American Medical Association*, volume 76, page 149, 1921); H. H. Dale: "The Nature and Cause of Wound Shock" (*Harvey Lectures*, 1919-1920, page 26; J. B. Lippincott Co., Philadelphia, publishers); F. X. Dercum: *Clinical Manual of Mental Diseases* (W. B. Saunders Co., Philadelphia).

An article by Cannon that reviews much of his work on the adrenals may be found in the *American Journal of Physiology*, volume 50, page 399, 1919; literature is appended. For articles by Stewart and Rogoff see, among others, the *American Journal of Physiology*, volume 46, page 89, 1918, volume 48, pages 22 and 397, 1919; *Journal of Pharmacology and Experimental Therapeutics*, volume 13, pages 95, 167, 183, 361, and 397, 1919, and volume

14, page 343, 1919; *American Journal of Physiology*, volume 51, page 366, 1920, and volume 52, page 304, 1920; *Journal of Pharmacology and Experimental Therapeutics*, volume 16, page 71, 1920, and volume 17, page 227, 1921; and the review by Stewart, "Adrenal Insufficiency" (*Endocrinology*, volume 5, page 283, 1921), which includes many references.

ORGANOTHERAPY. Consult Osborne's *Therapeutics* (W. B. Saunders Co., Philadelphia), and H. R. Harrower's *Practical Hormone Therapy* (P. B. Hoeber, New York). The *Wilson Laboratories*, Chicago, Ill., publish a quarterly, *The Autacoid and Suture*, which includes articles on the subject. The dangers involved in the use of glandular extracts is pointed out by M. P. Rucker and C. C. Haskell (*Journal of the American Medical Association*, volume 76, page 1390, 1921).

PLANT HORMONES. The literature on the subject is very meager. See, for example, J. Loeb: "Hormones in Bryophyllum" (*Science*, volume 44, page 210, 1916) ; E. J. Russell: *Soil Conditions and Plant Growth* (Longmans, Green & Co., London) ; R. W. Thatcher: *The Chemistry of Plant Life* (McGraw-Hill, New York).

A FEW CLASSICAL BOOKS AND PAPERS

Johannes Müller: *Lehrbuch der Physiologie*, volume 1 (Koblenz, 1844).

"Müller points out that the process of secretion consists of two phases,—the production of certain materials, and the casting out of these materials upon a surface either in the interior or upon the exterior of the body. The first phase he called 'secretion,' the second, 'excretion.' "

A. A. Berthold: "Transplantation der Hoden" (*Archiv für Anatomie und Physiologie*, page 42, 1849).
Berthold removed the testicles from cocks and grafted them to other parts of the body. He observed that "the animals retained their male characteristics in regard to voice, reproductive instinct, fighting spirit, and growth of comb and wattles."

Thomas Addison: *On the Constitutional and Local Effects of the Disease of the Suprarenal Bodies* (London, 1855).
An account of the now well-known "Addison's Disease."

Claude Bernard: *Leçons sur les propriétés physiologiques et les altérations pathologiques des liquides de l'organisme* (Baillière et Fils, Paris, 1859).
Here we find the first clear accounts of glandular organs that distribute their products by means of the blood stream. The very name "internal secretion" is due to this illustrious Frenchman.

Theodore Kocher: "Ueber Kropfexstirpation und ihre Folgen" (*Archiv für klinische Chirurgie*, volume 29, 1883).
Myxedema is due to the loss of the functional activities of the thyroid gland.

M. Schiff: "Bericht über eine Versuchsreihe betr. die Wirkungen d. Exstirpation der Schildrüse" (*Archiv für experimentelle Pathologie und Pharmakologie*, volume 18, 1884).

Epoch-making experiments on the effects of the removal of the thyroid.

P. J. Möbius: *Schildrüse theorie* (Schmidts Jahrbücher, volume 210, page 237, 1886).
The opinion is expressed that Basedow's disease depends on an abnormally increased activity of a ductless gland.

P. Marie: "Sur deux cas d'acromégalie, hypertrophie singulière non congénitale des extrémités supérieures, inférieures et céphaliques (*Revue de médecine*, page 298, 1886).
The discovery that acromegaly is a disease due to the pituitary.

Brown-Séquard: "Des effets produits chez l'homme par des injections sous-cutanées d'un liquide retiré des testicules frais de cobaye et de chien" (*Comptes rendus de la société de biologie*, volume 41, page 415, 1889).
"Brown-Séquard injected the juice of the testicle subcutaneously into his own body and observed an increase in corporeal and mental powers that he attributed to the influence of these injections." The beginning of modern organotherapy, or treatment by means of glandular extracts.

J. von Mering and O. Minkovski: "Diabetes mellitus nach Pankreas exstirpation" (*Archiv für experimentelle Pathologie*, volume 26, page 371, 1889).
The complete removal of the pancreas results in severe diabetes.

E. Gley: "Sur les effets de l'extirpation du corps thyroïde" (*Comptes rendus de la société de biologie,* page 843, 1891).
Tetany is due to the removal of the parathyroids.

G. Oliver and E. A. Schafer: "The Physiological Effects of Extracts of the Suprarenal Capsules" (*Journal of Physiology,* volume 18, page 230, 1895).
An injection of an extract of the adrenal glands increases the blood pressure.

E. Baumann: "Ueber das normale Vorkommen von Jod im Tierkörper" (*Zeitschrift für physiologische Chemie,* volume 21, page 319, 1896).
The discovery that the element iodine is a normal constituent of the body.

J. Takamine: "The Isolation of the Active Principle of the Suprarenal Gland" (*Proceedings of the Physiological Society,* in the *Journal of Physiology,* volume 27, page xxix, 1901).
An account of the isolation of adrenaline.

A. Fröhlich: "Fall von Tumor der Hypophysis cerebri ohne Akromegalie" (*Wiener klinische Rundschau,* 1901).
A disease is described which is the reverse of acromegaly—that is, it is due to a diminished activity of the pituitary.

W. M. Bayliss and E. H. Starling; "The Mechanism of Pancreatic Secretion" (*Journal of Physiology,* volume 28, page 325, 1902).
The hormone, secretin, is described.

J. S. Edkins: "The Chemical Mechanism of Gastric Secretion" (*Journal of Physiology*, volume 34, page 133, 1906).
Extracts of the lining of the stomach cause an increased formation of gastric juice.

H. Cushing: *The Pituitary Body and Its Disorders* (J. B. Lippincott Co., Philadelphia, 1912).
A classic on the subject.

E. C. Kendall: "The Isolation of the Iodine Compound Which Occurs in the Thyroid" (*Journal of Biological Chemistry*, volume 39, page 125, 1919).
The isolation of "thyroxin," the active principle of the thyroid gland.

INDEX

213

www.ingramcontent.com/pod-product-compliance
Lightning Source LLC
Chambersburg PA
CBHW071414170526
45165CB00001B/269